+ARCHITECT 02

STUDIO Pei-Zhu
STUDIO MAD

Inside Beijing Now

Futuristic Architecture in the age of Experimental Architecture

실험주의 건축 시대의 미래주의 건축 Shi Jian 스 지엔

스 지엔은 IS리딩 컬쳐의 기획담당자로 오랫동안 도시건축에 대한 비평 및 연구를 해오면서 도서 편집, 출판, 전시 분야에서도 활동해왔다. 그는 시티 차이나(City China)의 컨설턴트이자 투데이스 뱅가드(Today's Vanguard)의 편집 위원회 회원이며, "재생 전략: 베이징의 구도시— 시쓰 베이 국제 초청 전시(베이징, 2007)"를 기획했으며, 2008년 뉴욕 건축 센터에서 열린 "빌딩 차이나: 다섯 개의 프로젝트, 다섯 개의 이야기"를 웨이 웨이 섀넌과 공동으로 진행했다.

Shi Jian is a Planning Director of ISreading Culture Co.,Ltd. He is an urban and architectural critic and researcher, and initiates book editing and publishing, and exhibition. He is the consultant of City China, and member of editorial committee of Today's Vanguard. He initiated "Regeneration Strategy: Beijing's old city--Xisi Bei International Invitation Exhibition"(Beijing,2007), and initiated the exhibition "Building China: Five Projects, Five Stories" in New York Architectural Center (co-operated with Wei Wei Shannon, New York, 2008)

중국의 실험주의 건축은 1980년대 문화산업과 '85 뉴 웨이브'라는 현대미술 운동에 힘입어 시작되었다.
하지만 사회적 빈곤 및 디자인을 지배해온 민족주의 양식으로 실험주의 건축은 그저 논의되는 수준에 머물렀다.
1996년, 미국에서 교육을 받은 장용허는 중국으로 돌아와
베이징 시슈 북하우스(Beijing Xishu Book House)를 완성시키며 중국 실험주의 건축의 탄생을 알렸다.
당시의 현대건축은 민족주의 양식과 차별화하고 디자인의 질을 높이고자 시도했던 근대건축 양식을 의미했다.
하지만 그와 같은 중국의 근대건축과 아방가르드 디자인은 빠른 경제성장과 도시화로 곧 위기에 처했다.
민족주의 양식은 구식이 되었고 다양한 새로운 건축적 형태를 요구하는 건축계에서 살아남기 위해
'신속히 디자인'하는 현상이 지배적으로 나타났다.
그 결과 실험주의 건축이 논하던 문제들은 곧 잊혀졌으며, 실험주의 건축가들이 주장해온 아방가르드적인 태도는
자국주의, 이론주의, 미래주의 등 3개의 새로운 흐름으로 나뉘었다.
문화적으로 역행하며 자국주의 건축가들은 변화하는 현실에 반응하기 위해 지역 문화를 자원으로 재활용했고,
더 많은 건축가가 동참하고 결실을 맺으면서 자국주의 건축은 실험주의 건축의 몰락 후 건축계의 주류를 이어갔다.
한편 주 페이, 마 옌송 등 긍정적인 현실과 미래의 이미지에 집중하고 있는 미래주의 건축가들은 자국주의 건축을 거부하며
미래주의 건축을 국제적으로 가장 영향력 있는 건축적 아이디어로 이끌고 있다.

미래주의 건축은 전통을 거부하는데, 전통을 무시하지는 않되 거리를 유지한다.
미래주의 건축은 빠르게 진행 중인 중국의 도시화를 긍정적인 시각으로 바라보며
세계적인 비전 아래 도시의 활성화를 시도한다.
미래주의 건축은 정보화시대의 건축으로 미디어와 정보에 강한 연계를 갖는다.
미래주의 건축은 행위주의 시대의 새로운 산물이다. 혁신적인 재료와 구조 기술로 목적을 달성한다.
미래주의 건축은 여러 분야와의 협력을 도모한다.
건축, 제품 디자인, 그래픽 디자인 간의 경계가 점점 모호해지고 있다.

Experimental architecture in China began to emerge in the 1980s, influenced by the cultural industry, especially the "85 New Wave" movement in contemporary art.
However, it manifested itself only on a theoretical level, due to social poverty and the constraints of a prevalent Nationalist design style.
In 1996, US-educated Chang Yungho returned to China and completed the Beijing Xishu Book House, which signified the birth of Chinese experimental architecture.
Contemporary architecture at that time referred to a modern architectural style, which differed from the prevalent Nationalist style, seeking for the quality in design.
However, this kind of modern architecture and avant-garde design, in a Chinese sense, was quickly challenged by rapid economic growth and expansive urbanization.
The Nationalist style was out, and the large demand of new architectural forms lead to a general phenomenon of "fast design," which was the basics to survive in the architectural field.
As a result, the problems that experimental architecture faced were quickly forgotten.
The avant-garde attitude that experimental architects used to claim were changed, and later diverted into three directions: the domestic, the academic, and the futuristic.
Moving backwards culturally, domestic architects began reusing local cultural resources to respond to the great changes in reality. Due to the large numbers of architects who began practicing it (and showed fruitful results), the emerging domestic became the mainstream, following the breakdown of experimental architecture.
On the other hand, futuristic architects such as Zhu Pei and Ma Yansong resisted the domestic, looking at the positive reality and imagery of the future, making futuristic architecture their most influential vision.

Futuristic architecture refused the traditional, yet not to disrespect it, rather but to stay away from it.
Futuristic architecture positively responds to reality of rapid urbanization in China.
It tries to re-vitalize the city with a global vision.
Futuristic architecture is the contemporary architecture in the age of information.
It possesses a strong connection to media and information.
Futuristic architecture is the newcomer in this performative age.
Revolutions in material and structural technology have made its arrival possible.
Futuristic architecture has a strong cross-disciplinary desire.
The boundary between architecture, product design, and graphic design is gradually blurring.

04 FUTURISTIC ARCHITECTURE IN THE AGE OF
EXPERIMENTAL ARCHITECTURE 실험주의 건축 시대의 미래주의 건축_스 지엔

16 WHAT IS THE MEANING OF CHINESE TO YOU? 우리에게 중국적인 것은 어떤 의미인가?_주 페이 + 마 옌송

studio Pei-Zhu

28 DIGITAL BEIJING 디지털 베이징

38 GUGGENHEIM ART PAVILION 구겐하임 아트 파빌리온

44 ART MUSEUM OF YUE MINJUN 위에 민쥔 미술관

48 PUBLISHING HOUSE 퍼블리싱 하우스

52 BLUR HOTEL 블러호텔

60 CAI GUOQIANG COURTYARD HOUSE RENOVATION 차이 궈치앙 전통 주택 리노베이션

66 BEIJING XISI BEI REGENERATION STRATEGY 베이징 시쓰 베이 재생 전략

70 A CHINESE PROPHECY ON FUTURISTIC ARCHITECTURE 미래주의 건축에 대한 중국의 예언_조우 롱

studio MAD

76 HONGLUO CLUB 홍루오 클럽
84 TOKYO GALLERY BTAP RENOVATION 도쿄 갤러리 BTAP 개축
88 THE ABSOLUTE TOWERS 앱솔루트 타워
92 CHANGSHA CULTURE PARK 창샤 문화 공원
95 ERDOS MUSEUM 에르도스 박물관
98 DENMARK PAVILION 덴마크 파빌리온
100 MONGOLIAN PRIVATE MEADOW CLUB 몽골리안 회원제 메도우 클럽
102 GUANGZHOU CLUBHOUSE 광저우 클럽하우스
104 AL ROSTAMINI HEADQUARTERS 알 로스타미니 본사
106 BEIJING 2050 베이징 2050
110 SINOSTEEL INTERNATIONAL PLAZA 시노스틸 인터내셔널 플라자
114 WATER TANK FOR GOLDFISH 금붕어 수조
116 TOKYO ISLAND 도쿄 아일랜드
118 XIAMEN MUSEUM 샤먼 박물관
122 WHY MAD IS MAD 왜 MAD는 열광적인가? _ 지앙 쥔

WHAT IS THE MEANING OF CHINESE TO YOU?
우리에게 중국적인 것은 어떤 의미인가?

Zhu Pei (Studio Pei-Zhu) + Ma Yansong (Studio MAD) 주 페이(스튜디오 페이주) + 마 옌송(스튜디오 매드)

당신들에게 '중국적인 것'의 의미는 무엇인가?

스튜디오 페이주 사실 중국의 건축가들은 수 세대에 걸쳐 그와 같은 질문을 받아왔다. 내게 '중국적인 것'은 문화, 철학, 역사를 아우르는 포괄적인 개념이다. 지난 세기 초를 돌이켜보면 중국의 건축가들은 중국적인 것에 대해 생각해보려 하지 않았다. 많은 세대가 중국 전통 미학의 재생산이라는 덫에 걸려들었다. 우리는 중국적인 것을 논하며 단지 전통을 반영하거나 참조하는 데 그쳤다. 오늘날 중국의 설계에서 정체성을 찾아보기는 어렵다. 모두가 정체성을 파악하기 위해 고군분투 중이며, '중국적인 것'을 표현할 언어를 확립해가고 있다. 젊은 건축가들 역시 '무엇이 중국적인 것인지'를 고민한다. 하지만 대부분의 건축가들이 버내큘러(vernacular) 건축이나 중국의 전통 건축에 집착할 뿐 현대적인 시각을 갖고 있지 않다.

내가 볼 때 이 문제는 중국의 젊은 건축가들이 넘어야 할 가장 큰 산이다. 앞서 나는 구세대의 설계 방식에 대해 지적했다. 그들이 하는 '중국식 혹은 중국적'이라는 말은 편견으로 얼룩져 있다. 그들은 중국 건축을 창조하기 위해 서양의 영향과 모더니즘을 받아들이려고 했다. 모더니즘은 중국의 문화와 완전히 융화될 수 없었다. 극단적인 편이 차라리 쉬웠다. 즉, 전통을 복제하거나 전통과의 관계를 무시해버리는 것이다. 철학과 문화를 보면, 중국에서 설계가 어떤 맥락에 놓여 있는지를 알 수 있다. 하지만 중국의 철학을 이용해 현대적인 건물을 세우고자 한다면, 스스로에게 '그래, 이게 바로 중국적인 거야'라고 말하는 우는 범하지 말아야 한다. 우리는 전통 건축의 복제품이 아닌 중국의 현대건축이 등장하기를 기대한다. 중국적인 것을 위해 과거만 되풀이해서는 안 된다.

전통적인 버내큘러 건축은 중국 현대건축의 진정한 희망이라고 할 수 없다. 우리가 중국적이면서 현대적인 건축을 실현할 수 있는 유일한 방법은 중국의 건축을 현대적인 관점에서 이해하는 것이다. 중국의 철학적 지식을 활용해 문화를 현대적으로 이끌어나간다면 변화의 전기를 마련할 수 있다. 중국의 역사적인 건물들은 경이롭기도 하고 인류에게는 가치 있는 문화유산이기도 하지만, 어쩔 수 없는 과거의 산물이다. 우리는 그러한 역사적인 건물에 손을 댈 수 없다. 왜냐하면 어떻게 손을 대야 할지 모르기 때문이다. 과거의 방법을 재현한다고 해도 그것은 현실적이지 못하다. 어차피 빠르게 변해가는 현대문명에 휩쓸릴 것이므로 부질없는 노력이 될 것이다.

나는 현재 우리가 하는 작업이 미래에는 전통이 된다고 믿는다. 오늘 당신의 작업이 내일의 전통을 만드는 것이다. 전통적인 건축물도 건축 당시에는 아방가르드 하거나 현대적이었다. 스튜디오 페이주는 중국 예술이나 전통 건축에서 무언가를 차용할 때의 위험성을 잘 알고 있다. 우리의 유일한 희망은 현대문화 속에 있다. 우리는 지혜로운 중국 철학의 힘을 빌려 세상에 기여할 수 있다.

스튜디오 매드 정말 어려운 질문이다. 중국은 오랜 역사와 문화를 지닌 곳이지만 매우 다원화된 데다 최근에는 개발 열풍에 휩싸여 있다. 이렇게 혼란스런 상황에서 그와 같은 질문에 대답하기는 굉장히 어렵다. 하지만 이 시기를 잘 헤쳐 나갈 수 있는 방법은 분명히 있다고 본다. 아직도 중국에서는 말 그대로 과거의 것들을 차용하는 행위가 광범위하게 자행되고 있다. 한마디로 전통적인 형태, 재료, 상징이 반복 재생산되고 있는 현실이다.

나는 제11회 베니스비엔날레에서 이제 막 돌아왔다. 일본관은 일본인들이 이미 자기 정체성을 찾았다는 것을 보여주었다. 그들의 정체성은 극도로 가볍고, 극도로 얇고, 극도로 깔끔한 것이다. 중국관은 지역성에 초점을 맞추고 벽돌, 나무 막대와 같이 매우 거칠고 중국적인 소재를 사용했다. 하지만 그러한 재료들이 사용된 방식은 너무나 서양적이었다. 나는 그것이 중국의 미래를 보여준다고 생각하지 않는다.

스튜디오 매드를 설립했을 때 나는 스물 아홉이었고, 서양에서 현대적인 건축 교육을 받고 돌아온 직후였다. 3~4년 전에 수행했던 프로젝트들을 돌이켜보니 우리가 배운 것을 제대로 써먹을 시간이 없었고, 그러다 보니 서양의 건축 사상과 유사한 부분들이 어느 정도 남아 있을 수밖에 없었다는 생각이 든다. 하지만 당시 작품들을 보면, 예를 들어 홍루오 클럽하우스 같은 경우 중국적인 색채가 조금씩 드러나기 시작했다는 것을 알 수 있다. 홍루오 클럽하우스에서 우리는 떠있는 지붕을 설계하고 지붕 가장자리 안쪽에 유리 벽을 설치해 반 실내 외 공간을 창출했다. 이곳에서는 중국 전통 정원 체험이 현대적으로 재현된다.

당신들의 중국 건축가로서의 '정체성 찾기 과정'에 대해 이야기 해달라.

스튜디오 페이주 위에 민쥔 미술관의 경우 우리는 주로 자연에서 영감을 얻었다. 나는 위에 민쥔 미술관이 자연 경관의 일부가 되길 바랐다. 하지만 우리는 이 건물에 토착 미학이나 전통적인 기술을 적용하지 않았다. 실제로는 최첨단 재료들을 이용해 건축했다. 이 건물은 토착적이거나 자연적이라고 할 수도 있지만 그와 동시에 매우 현대적이다. 우리는 '어떻게 하면 현대적이면서도 자연의 일부가 되는 건물을 설계할 수 있을까?' 고민했다.

사람들은 디지털 베이징을 보고 정보 올림픽을 위한 관제센터라고 생각했을 것이다.

What is the meaning of "Chinese" to you?

studio Pei-Zhu This question actually carries through many generations of Chinese architects. For me, the meaning of "Chinese" is an integration of culture, philosophy, and history. If we go back to the early part of the last century, Chinese architects did not look for ideas related to the Chinese. For many generations they fell into the trap of reproducing traditional Chinese aesthetics. When we talk about Chinese, it is looks to only reflect or bear influence from the tradition.

Contemporary Chinese design has little if any identity today. Everybody is trying very hard to define this identity, establishing a language that reflects "Chinese."

Even today, for the young Chinese architect, it is still a struggle to find what the meaning of Chinese is. But most of the architects fall into the vernacular or traditional Chinese architecture, not a contemporary perspective.

I believe it to be one of the biggest challenges for young Chinese architects. I have pointed out the older generations' approach to design. When they mention "Chinese style, Chinese character," they are basing their ideas on a biased system. They want to use western influences, modernism, to create Chinese architecture. Modernism is not capable of completely blending with the Chinese culture. The easy route is to follow extremes; either replicating the traditional or ignoring the connection with culture altogether. Philosophy and culture can be great devices to view design's connection with its context in China. But, even when you use Chinese philosophy to create a contemporary building, do not fall into the trap of saying to yourself, "okay, this is Chinese." We can look forward to the Chinese architecture which is contemporary, no longer a repetition of the past. Repeating the historic character in an attempt to be Chinese is wrong.

Traditional vernacular construction is not the real hope for the Chinese contemporary architecture. The only way we can achieve both Chinese and contemporary is to understand it in a modern light. Using the knowledge of Chinese philosophy to carry the culture into a contemporary context becomes the path of change. Historical Chinese buildings are wonderful, and the most valuable things in the world, but they still belong to the past. We cannot work like that any more, since we do not know how to work on the traditional. Even when you work in a traditional way, this is not realistic. It becomes a futile effort, since it too will be swept away by the fast-paced contemporary culture.

I believe the contemporary work we are doing today is going to be traditional in future. Today's works are your contribution for the future's tradition. All the traditional architecture was avant-garde or contemporary at that time. Studio Pei-Zhu realizes the danger of borrowing something from the tradition of Chinese art or Chinese architecture. The only hope for us is on the contemporary culture. By using some smart Chinese philosophy, we can contribute something to this world.

studio MAD I think it's a really tough question, as China is so diverse, with such a long history of culture and is under such a crazy development today. It is a huge task to look for an answer in this chaos. But I think there must be a way we can bring this huge history to our future. In China right now, however, we are still largely at the stage of literally taking from history: copying traditional forms, materials, and symbols.

I have just returned from the 11th Venice Biennale. From the Japan Pavilion, I can tell that they already have their identity; it is super light, super thin, and super neat. The China Pavilion focused on local, Chinese materials, very rough materials like bricks and wooden bars. Yet these materials were still put together in quite a Western way. I don't think that is the future for China.

I founded Studio MAD when I was 29, after I returned from a modern architectural education in the west. Looking back on our projects of 3 or 4 years ago, it seems there was no time to reflect on what we had learnt, so there are still some similarities to western architectural ideas. But I think we can start to see some Chinese characteristics in our work of that time. Like [Hongluo Clubhouse](#), for instance. Here, we designed a floating roof with glazing set back from the roof edge to create the semi-indoor/outdoor space. This is a modern experience close to what you could feel in the traditional Chinese garden.

Tell us your identity-searching progress as a Chinese architect.

studio Pei-Zhu In projects like [Yue Minjun Museum](#), nature becomes our major influence. I hope this museum can be a part of the landscape. At the same time, we did not use vernacular aesthetics or traditional technology to build this museum. Actually, we used very high-tech materials. The building can be more vernacular or more natural, but the attitude can be more contemporary. We thought, "How we can produce something contemporary that belongs to nature?"

In projects like [Digital Beijing](#), people might think that this is a control centre for the Information Olympics. This is supposed to be built with a focus on high

중국의 몇몇 젊은 건축가는 여전히 전통적인 방식을 고수한다.
또 어떤 건축가들은 중국 문화와 어떤 식의 연관성도 없는 작품을
설계한다. 이들은 자신들의 설계와 작업 방식에 신경 쓰지 않지만
그들의 작품은 전통의 그림자 속에 있다. 특히 후자에 속하는
건축가들은 중국식이라기보다 서양식이라고 불러야 좋을 작품들을
생산한다. 실제로 중국 문화와의 연관성을 찾아볼 수 없기 때문이다.

디지털 베이징은 최첨단 기술에 초점을 맞추고 금속이나 유리를 재료로 건립했다. 기술 역시 자연에서 비롯된다는 것이 나의 견해다. 현대의 기술은 모두 자연에서 영감을 얻었다고 볼 수 있다. 우리는 로테크(low-tech)적이고 자연적인 방법을 사용해 하이테크적인 컨텐츠를 생산한다. 외관은 아무 의미가 없다. 다만 정체성과 태도를 분명히 하고 싶었을 뿐이다. 우리는 우리가 다르다고 생각한다. 우리는 '현대문화, 자연, 전통은 무엇인가'에 대해 차별화된 견해를 갖고 있다. 스튜디오 페이주의 모든 작업은 전통적인 것이라고 여겨지는 설계 미학을 따르지 않는다. 우리는 과거와의 연결고리를 찾으려 노력한다. 스튜디오 페이주가 특별한 이유는 바로 이 때문이다.

중국의 몇몇 젊은 건축가는 여전히 전통적인 방식을 고수한다. 또 어떤 건축가들은 중국 문화와 어떤 식의 연관성도 없는 작품을 설계한다. 이들은 자신들의 설계와 작업 방식에 신경 쓰지 않지만 그들의 작품은 전통의 그림자 속에 있다. 특히 후자에 속하는 건축가들은 중국식이라기보다 서양식이라고 불러야 좋을 작품들을 생산한다. 실제로 중국 문화와의 연관성을 찾아볼 수 없기 때문이다. 그러한 작품들은 중국의 과거 혹은 현재에 속하지 않는다. 사실 이 두 가지 방향은 모두 위험성을 내포하는데, 중국적인 것을 탐구하는 데 잘못된 길이 될 수도 있기 때문이다. 중국적인 것을 위해서는 중국적인 사고, 철학, 문화를 지녀야 한다.

스튜디오 매드 기본적으로 인간과 자연의 관계는 중국인들이 세상을 바라보는 방식에서 핵심을 이룬다. 우리 몸이 느끼는 감정은 중국 고대 건물에서 갖을수 있는 공간적 경험과 실제로 밀접하게 연관되어 있다. 우리 몸이나 자연에는 사각형이나 상자 모양이 존재하지 않지만 자연은 매우 강하다. 나는 우리가 자연의 법칙에 순응하면서 보다 유동적이고 유연하게 살아야 한다고 생각한다. 바로 이러한 개념이 토론토에 있는 앱솔루트 타워 설계의 밑바탕이 되었다. 우리는 단순한 수평 판을 우리 몸과 유사한 곡선 형태로 제작했다. 그러한 형태 또한 중국적인 정체성을 구성하는 요소 가운데 하나다. 나는 이러한 요소들을 탐구하고 종합하는 작업을 계속한다면 우리가 소위 말하는 중국적인 것의 정체를 밝힐 수 있다고 생각한다.

우리는 베니스비엔날레 국제관에서 '슈퍼스타'라는 명칭의 프로젝트를 전시했다. 우리는 무언가 위태로운 것을 만들어 역동적이고 변화무쌍한 중국의 기운을 보여주고 그것을 서양의 도시 속에 삽입하고자 했다. 그것은 선과 악을 오가는 모호한 것으로 전 세계 모든 도시에 약이 될 수도, 독이 될 수도 있는 것이었다. 우리는 세상 어디서나 발견할 수 있는, 보편적이지만 중국적인 내용을 담은 무언가를 만들고자 했다. 유명한 건축가들을 보면 국적이 어딘지 파악하기 힘들다. 그들의 작품은 사적이며, 예술 혹은 시대적인 개념과 관련이 있다. 그들은 국가가 아닌 개인의 정신세계를 표현한다.

당신들에게 중국적인 사고 방식이란 정확히 무엇인가.

스튜디오 페이주 중국 철학은 서양 철학과 상당히 다르다. 중국인들은 모든 것을 세계 또는 자연이라는 하나의 거대한 시스템 속에서 파악하는 반면, 서양 철학은 분석이나 시간에 따른 논리를 바탕으로 한다. 서양의 사고방식에 따르면 정보기술은 산업혁명에서 비롯된 것이다. 하지만 중국인들은 그런 식으로 사고하지 않는다. 중국인들에게는 만물이 순차적이지 않다. 이 세상에는 시작도 끝도 없으며 모든 것이 서로 맞물려 있다. 중국인들은 어떤 것을 대하든 스타일에 별로 신경 쓰지 않는다. 우리 인간은 돌이나 나무와 같은 존재다. 인간 역시 자연의 일부일 뿐이다. 반면 서양의 사상은 인간과 자연을 구분한다. 즉, 인간은 생태계에 속하고 돌은 물질계에 속한다. 만물은 독립적이며 논리와 시간이 사고의 근간을 이룬다.

스튜디오 매드 중요한 것은 재료가 아니라 '중국적인 방식'을 찾는 것이다. 대나무나 나무를 사용했다고 해서 그 건물이 중국 문화와 관련이 있다고 할 수는 없다. 나는 대나무를 좋아하지만 그렇다고 중국인들이 팬다는 아니다. 우리는 전통적인 상징이나 재료로 새로운 건물을 표현하는 일에는 흥미가 없다. 그게 중국적인 건축은 아니다. 중요한 것은 중국의 전통 건물에 내재된 생각들을 연구하는 것이다. 그러면 새로운 방식을 찾을 수 있을 것이다.

이제 우리는 우리의 정체성을 찾는 여정에 돌입했다. 지난 3년간 우리는 우리를 둘러싼 이 사회와 동일한 속력을 내며 달려왔다. 하지만 내가 볼 때 문화 축적은 속도가 느릴 때 더 잘 이루어진다. 다행히 나는 이제 속도를 줄일 수 있는 상황이다. 적어도 몇몇 프로젝트의 진행 과정을 적절한 속도로 조절할 수 있다.

베이징이나 당신들의 작품들 가운데 진짜 중국적이라고 생각되는 것이 있나?

스튜디오 페이주 나는 내 작품이 100% 중국적이라고 생각하지는 않지만, 매 작품마다 중국과 연관 지으려 노력한다. 내 목표는 모든 작품이 중국의 철학 및 문화와 관련된 것으로 비춰지게 만드는 것이다. 지난 작품들 중에는 만족스럽지 못한 것들도 더러 있지만, 중국의

In China some young architects still follow the traditional way. But others are doing something without any connection with Chinese culture. These people do not care which way they are working and their designs are under the shadow of tradition, whilst the second group of people have more similarity with Western style than Chinese. Actually it does not work within the Chinese culture.

technology, using materials such as metal or glass. For me, technology comes from nature. All the inspiration of modern technology may relate to nature. We use very low-tech and a very natural way for creating high technology contents. The shell does not mean anything. We only wanted to define a certain identity and attitude. We think we are different. We have a different judgment on "what is contemporary culture, nature, or tradition." None of my studio's work follows a traditional-looking design aesthetic. We try hard to look for some connection with the past, and this makes Studio Pei-Zhu quite unique.

In China some young architects still follow the traditional way. But others are doing something without any connection with Chinese culture. These people do not care which way they are working and their designs are under the shadow of tradition, whilst the second group of people have more similarity with Western style than Chinese. Actually it does not work within the Chinese culture. It does not belong to the Chinese past or the Chinese present. These two directions are dangerous. They become incorrect paths to explore something that belongs to the Chinese.

You have to have Chinese way of thinking, philosophy, and culture to have work that belongs to Chinese.

studio MAD Basically, the relationship between the human and the nature is the key point of the Chinese way of looking into the world. The physical feeling of the body is actually very closely connected with the spatial experience of ancient Chinese buildings. There is no square or box in our body, neither is there in nature, yet nature is so powerful. That makes me believe that we should follow the natural rule and to be more fluid and more flexible. That is the concept of our Absolute Towers in Toronto. Using a simple horizontal plate, we created the curved shape similar to the uniqueness of our own body. That is one element of our Chinese identity. And I believe if we keep exploring and collecting these elements, we will be able to find a so-called Chinese identity.

We exhibited a project called Superstar at the international pavilion in Venice Biennale. We were trying to make something dangerous, to represent the dynamic, ever-changing energy of China, and to inject it into Western cities. It was something equivocal, something between good and evil, between cure and poison for cities all over the world. We wanted to create something universal that could be found anywhere, but with Chinese contents inside. When you look at famous architects, you hardly find any national identity. Their work is personal, related to some art or concept from different periods. They present their own mind, not the country.

Tell us more about what exactly Chinese way of thinking is for you.

studio Pei-Zhu Chinese philosophy is very different from Western philosophy. Chinese people believe everything is contained in one system, which is this world or the nature, but Western philosophy is based on analysis, or logic based on time. Western thought draws the conclusion that information technology is produced from the industrial evolution.

But Chinese never think in this way. For Chinese people, not everything is sequential. This world has no ending point or starting point. It is pretty much interlocked together. Chinese people do not care about style when they deal with something. People like us are the same as a stone or a tree. Humans are only a part of nature. On the other hand, Western thought divides humans and nature; humans belong to ecology and stones belong to the physical world. Everything is independent, based on logic and time.

studio MAD Finding the 'Chinese way' is what's important, not the material itself. When you use bamboo or wood, that does not mean you are making a building that relates to your culture. I like bamboo, but Chinese people are not pandas. We are not interested in representing traditional symbols and materials in new buildings: this is not Chinese architecture. What's important is studying the ideas behind traditional Chinese buildings. Maybe then you can find a new way.

We are on the way now to look for our identity. In the past three years, we have been running at the same speed as society around us. But I believe culture is best accumulated in a slow way. Luckily, I have a better situation to slow down now; at least I can control a few projects with a proper development process.

Is there anything you think that is really Chinese in Beijing or in your work?

studio Pei-Zhu I do not think my work is fully Chinese, but we strive to find that connection in every project. Each project looks to be connected with Chinese philosophy and culture. That is my goal. Though I am not happy with some of my previous work, at least there is a continuous effort and strong intention to create a contemporary Chinese architecture. That is beyond my work, and you will understand my intentions by investigating my work. In my work, it does not matter what the form or the presentation is, one thing that is always beyond my work is strong attention to make connection with Chinese culture. This is pretty much consistent. Also, it is not really in a Chinese

이 자리에 새로운 건물을 배치하는 이유는 이미 그곳에 존재하던 것들을 무색하게 하기 위해서가 아니라 옛 것과 새것이 공존할 수 있다는 것을 보여주기 위해서다. 오래된 건물과 새로운 건물은 사람이 중심이 되는 빈 공간을 창출하기 때문이다. 나는 바로 이러한 관계가 전통적인 것이라고 생각한다.

현대건축에 기여하려는 지속적인 노력과 강한 의지만큼은 공통적이다. 그러한 노력과 의지는 작품을 초월한다. 내 작품들을 보면 무슨 말인지 이해할 수 있을 것이다. 내 작품에서 형식이나 건물의 외관은 별로 중요하지 않다. 하지만 모든 작품에 중국 문화와의 연결고리를 찾으려는 강한 의지가 매우 일관되게 반영되어 있다. 그리고 그러한 연결고리는 중국의 전통적인 방식에서 찾지 않았다. 내 작품은 현대와 중국과의 대화를 가장 지혜로운 방향으로 이끌어낸다. 베이징 구겐하임 미술관 혹은 차이 궈치앙 전통 주택 리노베이션과 같은 프로젝트에서 우리는 중국 전통 건축의 형식을 빌려왔다. 예컨대 차이 궈치앙 전통주택 리노베이션을 진행하며 나는 미니멀하고 미래적인 상자를 도입했다. 하지만 원래는 중국 전통 건축의 유형학을 참조해서 만든 것이다. 중국 전통 건축의 유형학에서는 중정, 마당, 그 밖의 공간을 구성하는 형식이 정해져 있다. 지붕은 관계 없으므로 제외시킬 수 있었다. 그렇다면 이는 무엇을 의미하는가? 중국에는 항상 문화나 위계질서를 드러내는 정해진 형식이 있었다.

베이징 구겐하임 미술관에서 우리는 버내큘러 건축의 비율과 척도를 참조해 '부유하는 상자(floating box)'를 만들었다. 마찬가지로 미술관의 유기적인 구조 역시 주변 환경에서 힌트를 얻었다. 우리는 중국의 방식을 존중한다. 중국의 전통적인 건축 설계에서는 정해진 형식은 물론 그에 따른 외부 공간도 중요하게 생각한다.

위에 민쿤 미술관은 도시가 아닌 한적한 자연 한가운데에 자리하고 있다. 주변이 산으로 둘러싸인 미술관의 형태는 커다란 바위 한 조각을 연상시킨다. 내 생각에 이런 점이 바로 중국적 사고의 영향인 것 같다. 우리는 미술관이 어떤 사물이 아니라 자연의 일부로 보이길 원했다. 이러한 작품들은 중국 철학 및 중국적 사고와의 끈을 놓지 않으려는 우리의 열정을 반영한다. 그것들은 우주에 대한 중국인의 관점을 드러내기도 한다. 이는 우리가 다른 이들과 크게 차별되는 부분이다. 요즘 사람들은 이전 세대보다 똑똑하지 않다. 우리는 어떻게 과거를 배울 수 있을까? 어떻게 과거의 정신적인 생각들을 우리의 설계에 접목시킬 수 있을까? 이러한 질문들이 현명한 해결책을 제시한다.

스튜디오 매드 나는 베이징에서 태어났기 때문에 베이징을 잘 안다. 베이징의 아름다움은 공공공간, 반 공공공간, 사적인 공간이 여러 겹으로 겹쳐 있다는 데 있다. 우리는 어릴 때 마당에서 논다. 하늘과 나무가 보이는 내적인 공간을 소유하는 것이다. 그런데 마당 밖으로 나가면 공공공간에서 많은 다른 아이가 함께 놀고 있다. 그곳에서 건물은 중요하지 않다. 건물은 단지 공공공간의 파사드를 형성할 뿐이다. 이렇게 겹쳐진 공간이 이 도시의 매력이다.

베이징 시내의 후통을 돌아다니다 보면 베이징 출신인 나조차도 잘 모르는 길들을 발견할 수 있다. 하지만 방향만 알면 목적지에는 확실히 도달하며, 길에서의 경험을 즐길 수도 있다. 가장 붐비는 후통에서도 즐겁게 돌아다닐 수 있다. 하지만 어떤 후통에서는 지나다니는 사람을 거의 보지 못할 수도 있다. 상당수의 후통은 주요 도로에서는 잘 보이지 않지만 최고의 음식점들과 갤러리들을 포함하고 있다. 후통의 이러한 요소들은 관광객들은 물론 부자들을 위한 가짜 신화를 창조하려는 개발업자들을 끌어들이지만, 사실 여기 사는 것은 굉장히 힘들다. 한 예로 이곳에는 개인 화장실이나 실내 화장실이 없다.

우리는 베이징 2050 프로젝트를 통해 후통에 자리한 몇몇 오래된 건물들을 유지하고, 그곳 거주민들을 위한 화장실이나 젊은 세대를 위한 소호 공간 같은 새로운 시설들을 제공하고자 한다. 이러한 제안과 함께 우리는 좀 이질적으로 보이는 버블을 기존 도시 조직에 도입하려고 한다. 마치 물방울처럼 각기 다른 형태를 띠고 있는 버블은 각자의 자리에 고정되고 기존 건물들과 연결된다.

이제 우리는 매드 사무실 근처 부지에 전통주택 개조 프로젝트를 시작하면서 우리의 생각들 중 일부를 실행에 옮길 수 있게 되었다. 전통주택 개조 프로젝트의 의뢰인은 베이징 2050을 보고 우리를 찾아와 "마음에 듭니다. 우리 같이 한번 해봅시다"라고 말했다. 의뢰인의 중정에는 2개의 건물이 자리하고 있었다. 하나는 진짜로 전통적인 건물이었고 다른 하나는 가짜였다. 우리는 가짜 건물을 철거하는 조건으로 진짜 건물을 개조하겠다고 했다. 현재 우리는 가짜 건물이 서 있던 자리에 버블을 건설하고 있다. 진정한 의미에서 현대적인 요소는 후통 지역 어디에도 존재하지 않는다. 우리는 주변의 오래된 건물들을 투영하는 유기적인 형태를 고안했다. 건물 안에는 화장실과 지붕으로 이어지는 계단이 있다. 이 자리에 새로운 건물을 배치하는 이유는 이미 그곳에 존재하던 것들을 무색하게 하기 위해서가 아니라 옛 것과 새것이 공존할 수 있다는 것을 보여주기 위해서다. 오래된 건물과 새로운 건물은 사람이 중심이 되는 빈 공간을 창출하기 때문이다. 나는 바로 이러한 관계가 전통적인 것이라고 생각한다.

이렇게 주변 상황을 반영하는 버블 프로젝트는 국립오페라하우스 류의 건물들과는 차원이 다르다. 국립오페라하우스는 첨단 재료들로 구성되었음에도 불구하고 현대적인 건물이 아니다. 주변에 대한 이 건물의 태도는 매우 서양적이다. 미국의 빌라와 마찬가지로 건물은 중앙에 자리하고 주변은 빈 공간이 감싸고 있다. 땅을 사서 그냥 가운데다 집을 지은 것 같다. 반면에 이웃에 자리한 자금성은 완전히 다른 토지 이용 방식을 보여준다. 이곳의 오래된 건물들은 뽐내기 위해서가 아니라, 도시 공간을 정의하고 건물 본연의 기능을 다하기 위해 설계되었다.

> **When you put a new building in this position, the goal is not to overshadow what's already there. Rather, it is to show our idea that the new and the old can work together, because both create an empty space where human is at the centre. I think it's this relationship that's traditional.**

traditional way. It is work that creates a dialogue between the contemporary and Chinese that becomes the most intelligent direction. In projects like the Guggenheim Museum in Beijing or the [Cai Guoqiang Courtyard House Renovation](), we abstracted form from the traditional Chinese architecture system. For example, in Cai Guoqiang Courtyard House Renovation, I introduced a box of minimalist and futuristic characteristics, but it actually draws from the traditional Chinese architectural typology. If you abstract Chinese architectural typology, they use regular forms to organize the courtyard, plaza, and the sequence of spaces. The roof becomes irrelevant, you can cut it out. What is then behind the language? The Chinese have always used the built form as a demonstration of their culture or hierarchy.

In Guggenheim Museum in Beijing, we used the proportion and scale of the vernacular architecture to influence our "floating box." Similarly, the museum's organizational structure takes its cues from the surrounding context. We respect the Chinese way. Traditionally, when you design a piece of Chinese architecture, you look not only at the built form, but also at the outdoor space created because of it.

The [Yue Minjun Museum]() is not a building in the city, but in an undisturbed natural surrounding. Influenced by the surrounding mountains, the museum reflects the form of a piece of rock. In my belief, this is the influence of Chinese thinking. Our intention is for the museum not to be viewed as an object, but as a part of nature. These projects strongly demonstrate our eagerness to create something related to Chinese philosophy, and thus Chinese thinking. It is a way to represent a Chinese perspective on the universe. That makes a huge difference for us. People now are not smarter than the previous generation. How can we learn from the past? How can we inject their spiritual ideas into our designs? That is a smart solution.

studio MAD I was born in Beijing and I know it very well. The beauty of Beijing City is layers, public layers, semi-public layers, and private layers. When you are a kid, you play in your own courtyard, you have this internal space with sky and trees. Once you go out of your own court, there are many other kids who can play together in public space. There, buildings are not as important, as the buildings are only the facade that forms the public space. This kind of layer is the beauty of this city.

You can go into the city centre Hutongs, and although I am from Beijing, even I do not know all of them. But if you know the direction you need to go in, you will definitely reach your destination, and enjoy the experience on the way. Even in the busiest Hutongs, you can totally enjoy yourself, whilst other times you might only see a couple of people. Many of the Hutongs actually contain the best restaurants and galleries, hidden away from the main roads. Yet, although they are a draw for tourists and for developers who want to create fake history for rich people, living in these Hutongs is actually quite hard: there are no private toilets or indoor bathrooms, for instance.

In our [Beijing 2050]() project, our idea is to keep some of the old buildings in the Hutongs, and to provide new interventions for people who live there: private toilets, or small office space for young people moving in. In this proposal, we added some alien-looking bubbles into the urban fabric. They are all different shapes and they are like water drops, dropped into each space to connect with the older buildings.

Now we have been commissioned to carry out some of our ideas, in a courtyard renovation project close to my office, a client saw Beijing 2050 and came to us, saying 'I like this, maybe we can try it'. The client's courtyard used to contain two buildings: one an authentic, traditional building, the other fake. We suggested we could renovate the traditional building, as long as we could tear down the fake one. So we are now building a bubble where the fake building stood. In the whole Hutong area there is no real contemporary element. As such, we designed an organic shape that reflects the old buildings around it. Inside, there is a toilet and stairs that go up to the roof. When you put a new building in this position, the goal is not to overshadow what's already there. Rather, it is to show our idea that the new and the old can work together, because both create an empty space where human is at the centre. I think it's this relationship that's traditional.

This reflective bubble project is totally different from something like the National Opera House. I do not think that this building is a contemporary building, although the materials are new. The way that building deals with its surroundings is very, very western. You put the building at the centre and leave all the space around it. It's like an American Villa. You buy the land and just put the house in the centre. In contrast, the Forbidden City, next door, is a completely different idea of land use. The ancient buildings here are designed to define the space of the city, and to work, not to show off themselves.

How do you think you can save the integrity of Beijing with its city fabric as an architect?

studio Pei-Zhu To demonstrate the idea of how to integrate or, to design into the existing urban fabric of this city, we designed work that is located in the

우리는 주변으로부터 완전히 독립적인 건물을 세울 수 없다. 우리는
환경과의 조화를 원하며, 그것이 꼭 전통으로의 회귀를 의미하지는
않는다. 우리는 '오늘날의 베이징' 혹은 '오늘날 베이징의 도시
환경'과 같은 주제와 우리의 설계를 연결하고 싶었다.

건축가로서 베이징이라는 도시의 통일성을 어떻게 유지할 수 있다고 보는가?

스튜디오 페이주 우리는 건물이 기존 도시 구조에 어떤 식으로 통합될 수 있는지를 보여주기 위해 베이징 시내 곳곳에 작품을 설계했다. 예컨대 어떤 작품은 구시가지 내 제2순환로에 있으며, 다른 작품은 제 2순환로와 1960년대에 소비에트 양식으로 건설된 제3순환로 사이에 자리한다. 신개발지구에 있는 디지털 베이징의 경우에도 우리는 건물이 기존 도시 구조에 융화되도록 최선을 다했다.
블러 호텔을 설계하면서 우리는 전통을 파괴하지도 전통으로 회귀하지도 않았으며, 주변의 전통주택에서 영감을 받아 중정을 수직적으로 배치하는 현대적인 접근 방식을 채택했다. 블러 호텔은 중국 전통 등(lantern)을 닮았으며 매우 동양적이다. 우리는 전통적인 방식으로 돌아가거나 완전히 낯선 건물을 세우지도 않았다. 우리는 현대적으로 접근해 우리 건물이 중국의 과거, 문화 그리고 주변 도시 구조와 조화되도록 했다.
퍼블리싱 하우스에서 우리는 건물을 철거하지 않고 기존 구조를 최대한 활용했으며, 이 건물이 쇠퇴해가고 있는 주변 지역에 일종의 자극제로 기능하길 원했다. 건물의 주변 지역은 매우 조용했으며 주거용 건물이 대부분이었다. 그렇기 때문에 퍼블리싱 하우스는 이 지역의 활성화에 상당히 기여할 것이다.
디지털 베이징의 경우, 베이징의 과거 또는 현재와 특별한 관련이 없다. 우리는 '중국의 오늘', '베이징의 오늘'을 보여주려고 했다. 이 건물은 다른 해외 건축가들과는 차별화된 우리만의 설계 방식을 강하게 암시한다. 우리는 주변으로부터 완전히 독립적인 건물을 세울 수 없다. 우리는 환경과의 조화를 원하며, 그것이 꼭 전통으로의 회귀를 의미하지는 않는다. 우리는 '오늘날의 베이징' 혹은 '오늘날 베이징의 도시 환경'과 같은 주제와 우리의 설계를 연결하고 싶었다. 디지털 베이징의 추상적인 형태 이면에는 중국적인 생각과 사고방식이 자리한다. 우리는 천연 재료를 사용했고, 외형에 중점을 두지 않았으며, 건물의 용도를 고려했고, 첨단 기술을 사용하지 않았다. 우리는 하이테크 건물에 로테크 방식을 적용하는 등 매우 자연적인 방식을 선택했다. 이 모든 노력이 오늘날의 베이징을 표현하기 위한 것이었다. 어떤 이들은 몇몇 건물을 보며 '베이징에 속하지 않는다'고 비판한다. 하지만 나는 그들이 틀렸다고 본다. 그들은 '그 건물들이 전통적인 베이징에 속하지 않는다'고 얘기한 것일 뿐이다. 이러한 현실에서도 우리는 중국인들의 사고방식을 엿볼 수 있다.
오늘날 중심업무지구의 풍경이 현대 중국의 진짜 모습이다. 건축가로서 우리의 역할은 사회 발전에 많은 영향을 미친다. 전통적인 중국식 건물에 들어가는 것도 하나의 경험이다. 우리 설계에 주로 영향을 주는 것은 바로 그와 같은 느낌이다.

우리는 어떤 비밀스런 방법이 아닌 논리와 중국적 사고방식을 기반으로 한다. 우리는 가짜 앤티크 건물을 양산하는 낡은 수법을 멀리한다. 우리는 현대적인 설계를 위해 추상적인 개념과 논리를 선호한다.

스튜디오 매드 구(舊) 베이징은 전통주택들로 구성되어 있다. 기본 요소가 되는 건물은 매우 작으며 반복적으로 밀집해 있다. 우리는 이를 '도시 조직'이라고 부른다. 현대사회에 가까워지면서 건물 규모는 점점 더 커졌다. 1950년대에는 공산주의 체제 10주년을 맞아 국립미술관과 같은 10개의 기념비적인 건물이 건립되었다.
현재 스튜디오 매드는 국립미술관의 정 반대편에 '이름 없는 마당(No-name yard)'이라는 새로운 프로젝트를 진행하고 있다. 이 지역은 매우 흥미롭다. 프로젝트 부지 바로 옆에는 여전히 전통주택들이 존재한다. 하지만 반대편에는 미술관을 비롯해 호텔, 쇼핑몰과 같이 높이가 30m에 이르는 현대적이고 상업적인 건물들이 늘어서 있다. 이 지역은 다양한 시대의 건물들이 만나고, 새로운 건물들이 오래된 도시를 야금야금 잠식해나가는 도시의 역사를 축약해서 보여준다. 우리 역시 작은 규모와 큰 규모가 공존하는 과도기를 지나고 있다. 우리의 부지는 중앙에 자리하는데 만일 이곳에 쇼핑몰과 같이 규모가 큰 건물이 들어서면 주변의 전통주택들은 살아남지 못하고 다른 곳으로 옮겨가야 할 것이다.
우리는 어떻게 하면 주변의 소규모 주택들과 공존하면서 개발업자가 원하는 거대한 볼륨을 창출할 수 있을지에 대해 고민했다. 그리고 그 결과 전체적인 볼륨은 거대하지만 수많은 부분으로 나뉘어 떠다니는 구름과 같은 건물을 고안해냈다. 이 건물은 마치 공중에 세운 수평적인 도시 조직과 같아서 뚜렷한 형태가 없으며, 유리 벽이 모두 뒤로 물러나 있어 공중에 떠 있는 슬래브만 눈에 띈다. 이렇게 실내와 실외의 경계가 모호한 탓에 지붕이 있는 외부 공간의 수는 더 많아졌다. 결국 우리는 '무형(無形)' 전략을 통해 도시 조직을 확장하고 있는 셈이다. 나는 그러한 점이 바로 구 도시의 매력이라고 생각한다. 구 도시는 실제로 아무도 계획하지 않기 때문에 매우 산만해 보이지만 알고 보면 무척 흥미롭다.

We did not build something totally independent from the surrounding. We wanted some integration, but it is not about going back to the tradition. We wanted to integrate our design with the subject of "What is today's Beijing".

---

different places of the urban Beijing. For example, some work inside of the old town, inside of the second ring road, or some in-between the second ring road and the third ring road, which was built in the 1960s, influenced by Soviet style.

And also, a project like Digital Beijing, which is located in very new developing zone, we worked hard to make our design integrated into the existing urban fabric.

In <u>Blur Hotel</u>, we did not simply demolish, nor went back to the traditional way; we took a very contemporary approach of using a vertical courtyard, which was inspired by the surrounding courtyard house. Blur Hotel resembles a Chinese lantern. It is very eastern that way. We did not go back to the traditional way, and did not build some strange contemporary object, either. We took a contemporary approach that makes our building connected to the past, to Chinese culture, and to the surrounding urban fabric.

In <u>Publishing House</u>, again, we did not demolish the old building. We took advantage of the existing structure. We intend this building to be a stimulating point, to have an influence on the surrounding declining area. It was a very quiet area and most of the buildings there are residential. So this building will play a significant role in stimulating this area.

In <u>Digital Beijing</u>, there is no specific connection with the present or the old Beijing. Our solution was to demonstrate "Today's China," or "Today's Beijing." Beyond this building, there is a strong implication or intention that makes our design quite different from other foreign architects' approaches. We did not build something totally independent from the surrounding. We wanted some integration, but it is not about going back to the tradition. We wanted to integrate our design with the subject of "What is today's Beijing," or "What is today's Beijing Urban Environment." Beyond this abstract form of Digital Beijing, we had Chinese considerations, or used a Chinese way of thinking; we used natural materials; we did not focus on the shell; we thought about the content of this building; and we did not use high technology to build this building. Our way was pretty much natural: low-tech methodology for a high-tech building. All the effort was to connect to today's Beijing. Some criticize some buildings saying, "Those do not belong to Beijing." But I think they are wrong. They are only saying "Those do not belong to the traditional Beijing." Behind this contemporary way of thinking, we have a Chinese way of thinking.

What you see in the CBD today is a true vision on contemporary China. As architects, we play an influential part in the development of our society. It is an experience to enter a traditional Chinese building. That feeling is a major influence on our designs.

We do not use secret elements; we use logic and a Chinese way of thinking. We do not want to fall into the old approach of reproducing fake antique buildings. We prefer to use abstraction and logic to approach modern design.

studio MAD Old Beijing city is composed of courtyard houses. The basic element, the building, is very small and repeated in dense scales. We call it 'urban fabric'. As you get closer to modernity, the scale gets bigger. In the 1950s, 10 huge, monumental buildings were built in the city, to celebrate ten years of Communism, like the National Art Museum.

Now, Studio MAD has a new project called 'No-name yard', exactly opposite the National Art Museum (a project we almost won). This area is very interesting. Next to our project site, there are still courtyards, but on the other sides, as well as the museum, are modern, commercial buildings: hotels and shopping malls, 30 metres tall. This area tells the history of the city in microcosm, a place where all the new buildings from different times meet, where all the new projects are trying to eat the old city, piece by piece. We are also at a transition point between small and large scale. Our site is at the centre. If we build a large scale building, like a shopping mall, those courtyards cannot live. They have to move.

Our challenge is how to extend the small scale into our site, whilst also achieving the large volume that the developers want. So we have created a building that's like a floating cloud, composed of many, many smaller pieces, but with a large volume overall. It is like a horizontal urban fabric, piled up in the air, with no clear building shape, and with all the glass set back, so you can only see the floating slabs. Thanks to this blur between indoor and outdoor, you have many covered outdoor spaces. So we are using the 'no-shape' strategy to extend the city fabric. I think that is the beauty of the old city. Nobody really drew a plan so it looks so messy, but the space is very interesting.

CONTEMPORARY/
RARY/
MODERN/
CONCEPTUAL/TODAY'S
BEIJING/
CHINESE
THINKING/
RECONNECT/
ATTENTION

STUDIO
Pei-Zhu

스튜디오 페이주는 베이징을 중심으로 활동하는, 30명 남짓으로 이루어진 젊은 건축설계사무소다. 우리에게 디자인은 혁신적인 개념적 사고와 비판적인 시각을 가지고 실용적인 해결책을 제시하는 것이고, 우리의 프로젝트는 과정에서 결과로 이어지는 방법을 탐구하고 있다.
우리 작업이 지닌 탐구적이고 실험적인 특징은 중국의 도시 환경이라는 맥락 안에서 이루어진다. 최근 중국에서 일고 있는 빠른 개발은 새롭고 현대적인 도시 환경을 만들고 있지만 옛 지역의 정신이나 활기는 없다. 우리 디자인이 주목하고 있는 주요 이슈 중 하나는 중국의 현대적인 도시를 그 뿌리에 재접목시키는 것이며, 도시에서 일어나는 여러 활동에 활력을 불어넣을 수 있는 건축적 장치를 만들기 위해 현대적인 컨텍스트 안에서 버내큘러의 재해석을 시도한다. 우리는 이러한 작업을 통해 지역 컨텍스트에 적합한 현대건축의 지역적 다양성에 기여하길 희망한다.

Studio Pei-Zhu is a young practice of under 30 employees, based in Beijing. For us, the challenge of design is to provide practical solutions while reflecting a strong and innovative conceptual thinking and a critical outlook. Our projects, therefore, are an exploration of methods to connect process to product.
The framework for this investigation and the experimental nature of our work is formed by the context in which it takes place: Urban China. The recent rapid development of the country has created new urban environments that can certainly be described as modern, but that lack the vitality and soul of older districts. One of our main concerns in design is to reconnect modern urban China to its roots, reinterpreting the vernacular in a contemporary context to create architectural devices capable of energizing urban activities. In this way we hope to contribute to a regional variance of contemporary architecture appropriate to its local context.

ZHU PEI 주 페이

주 페이는 스튜디오 페이주의 대표이자 창립자로 칭화대에서 건축 석사를, UC버클리에서 건축과 도시 디자인 석사 학위를 받았으며, 미국 설계사무소의 프로젝트 디자이너 등으로 10년 이상 실무 경험을 쌓았다. 유네스코가 수상하는 'Special Merit 상', 중국 건축예술상(China Architectural Arts Award), 아키텍추럴 레코드 사의 '차이나 어워드', '디자인 뱅가드'상을 수상했다. 파리, 바르셀로나, 런던, 로테르담 등 세계 각지에서 '중국 현대건축' 관련 전시에 참여했고 뉴욕타임스, 르몽드, 타임, 가디언, LA타임스, 도무스, 아키텍추럴 레코드, 월 페이퍼 등의 매체에 인터뷰와 작품이 게재된 바 있다. 또 다른 대표인 우 통와 함께 세계적인 설계회사를 대상으로 한 지정 공모전이었던 베이징올림픽 관제센터 공모전에서 우승함으로써 베이징올림픽을 위한 3개의 주 건축물(베이징올림픽 주경기장, 베이징올림픽 수영장, 베이징올림픽 관제센터) 중 하나인 '디지털 베이징'을 디자인했으며, 구겐하임 재단 선정으로 베이징 구겐하임 미술관과 프랭크 게리, 자하 하디드, 장 누벨, 안도 타다오 등 세계적인 건축가들이 참여하는 13개 아부다비의 구겐하임 아트 파빌리온 프로젝트 중 하나를 설계하고 있다.

WU TONG 우 통

우 통은 스튜디오 페이주의 대표(design principal)로 2003년부터 현재까지 '예술과 디자인', '디자인' 지의 아트 디렉터로 활동하고 있다. 칭화대에서 예술 디자인 석사 학위를 받았고, 2000년에 홍콩 디자이너협회의 우수상을 수상했으며, 2006년에 베이징 국립박물관의 국제 디자인 전시에 참여했다.

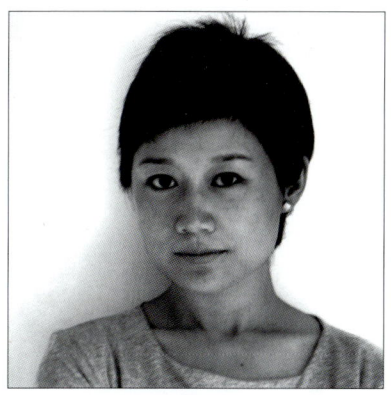

Zhu Pei is the president and founder of Studio Pei-Zhu. He built his career as a project designer, having spent more than 10 years at architectural design firms in the United States. He completed a Master of Architecture degree at Tsinghua University and a Master of Architecture and Urban Design degree at UC Berkeley. He received the Award of Special Merit from UNESCO, the China Architectural Arts Award, the China Award from Architectural Record, and the Design Vanguard from Architectural Record. He has participated in exhibitions related to China's contemporary architecture across the globe, including Paris, Barcelona, London, and Rotterdam. His interviews and works have been featured in various media, including the New York Times, Le Monde, The Times, The Guardian, LA Times, Domus, Architectural Record and Wallpaper. Along with Wu Tong, design principal of the studio, he has won the first prize in the Beijing Olympics Control and Data Centre competition, a designated competition for world-class design firms; he thus designed the "Digital Beijing", one of the three main buildings for the Beijing Olympics (the Beijing Olympics main stadium, the Beijing Olympics swimming pool, and the Beijing Olympics Control and Data Centre). Selected by the Guggenheim Foundation, he is currently designing the Guggenheim Museum Beijing and one of 13 Guggenheim Art Pavilion Projects in Abu Dhabi, in which world-renowned architects are participating, including Frank Ghery, Zaha Hadid, Jean Nouvel, and Ando Tadao.

Wu Tong is a design principal at Studio Pei-Zhu, and has been the art director of "Art and Design" and "Design" from 2003 to the present. She received a Master's degree from the Academy of Art Design at Tsinghua University, and she was the recipient of the Excellent Award from the Hong Kong Designers Association in 2000. She also participated in the International Design Exhibitions, featuring the National Museum of China in 2006.

DIGITAL BEIJING

디 지 털 베 이 징

Beijing Olympic Park, Beijing, China

Architects: Studio Pei-Zhu & Urbanus
Project designer: Zhu Pei, Wu Tong, Wang Hui
Project team: Liu Wentian, Li Chuen, Lin Lin, Tian Qi
Program: Control and data centre of Beijing Olympics
Structure: Reinforced concrete and steel frame
Floor: B2, 11F
Site area: 16,018m^2
Total building area: 96,518m^2
Engineer: China Institute Building Standard Design & Research
Design period: 2004-2005
Construction period: 2005-2008
Client: Beijing Network Information Industry Office

베이징 시정부는 2008년 올림픽 역사상 가장 화려한 기술 콘텐츠를 담은 올림픽을 선보이겠다고 다짐했다. 디지털 올림픽의 랜드마크인 디지털 베이징 건물은 국립경기장과 국립수영센터가 자리한 올림픽센터를 관통하는 중심축 북쪽 끝에 있다. 우리의 디지털 베이징 디자인은 국제적인 명성을 얻은 7개의 건축설계사무소가 참여한 설계경기에서 당선되었다. 100,000m^2 면적에 달하는 디지털 베이징은 2008 베이징올림픽 기간 동안 통제센터 및 데이터센터로 기능했으며, 추후에는 디지털 올림픽 가상 뮤지엄과 디지털 제품 생산업체를 위한 전시시설로 기능할 것이다. 빠르게 진화하는 디지털 시대는 우리의 삶, 사회 그리고 도시에 큰 영향을 주고 있다. 근대주의가 산업혁명의 결과였다면 디지털 베이징은 디지털 시대의 가능성을 시험하고 있다고 할 수 있다. 디지털 베이징의 개념은 숫자와 정보가 지배하는 현시대의 건축을 다시 고려하고 재조명하며 발전시켰다. 디지털 베이징은 고요한 수면 위로 디지털 바코드와 집적 회로기판을 연상시키는 모습을 드러낸다. 우리는 구체적인 형태를 원했으며, 우리 삶을 가득 메우고 있지만 쉽게 간과되는 마이크로칩을 연상시키듯 아주 작은 세계를 확대해서 보여주고 싶었다. 숫자 0과 1의 단순한 반복 같은 추상적인 매스를 통해 디지털 올림픽과 정보시대의 강렬한 상징이 될 것이다. 디지털 베이징은 진화하는 기술에 맞춰 미래에도 계속해서 변화할 것으로 기대한다.

The Beijing Municipal Government had promised the world that Beijing would present an Olympics in 2008 that will have the highest technological content in history. As the landmark building of the Digital Olympics, the Digital Beijing Building is located at the northern end of the central axis, neighboring the core area of the Olympics Centre, the National Stadium, and National Swimming Centre. Seven internationally renowned architectural firms participated in the competition for Digital Beijing, and our scheme was selected a winning scheme. The Digital Beijing Building, nearly 100,000 square meters in area, served as the control centre and the data centre of the Digital Beijing Olympics, 2008. At other times, it will accommodate a virtual museum of Digital Olympics, and an exhibition centre for manufacturers of digital products. The rapid development of the digital age has greatly impacted our society, life, and city. If the industrial revolution has resulted in Modernism, the Digital Beijing Building begins to explore what will occur in the digital epoch. The concept for Digital Beijing was developed through reconsideration and reflection on contemporary architecture in the digital and information era. Appearing like a digital bar code and an integrated circuit board, the building emerges from a serene water surface. Here we are seeking a specific form, trying to reveal an enlarged micro world, suggestive of the microchips that are abundant but ignored in our daily lives. With an abstracted mass, as in the simple repetition of the digits 0 and 1, this building will be an impressive symbol of the Digital Olympics and of the information era. In the future, we expect that the Digital Beijing Building will always be under renovation, evolving and keeping pace with technology.

Photographs by Iwan Baan

1. Beijing Olympics control and data centre
2. Beijing Olympics swimming pool
3. Beijing Olympics main stadium

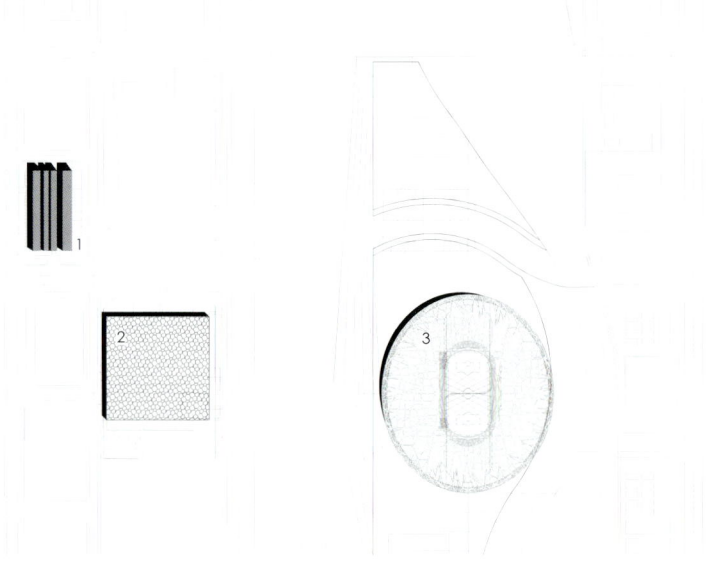

Site plan

APPEARING LIKE A DIGITAL BAR CODE AND AN INTEGRATED CIRCUIT BOARD, THE BUILDING EMERGES FROM A SERENE WATER SURFACE.
디지털 베이징은 고요한 수면 위로 디지털 바코드와 집적 회로기판을 연상시키는 모습을 드러낸다.

AT OTHER TIMES, IT WILL ACCOMMODATE A VIRTUAL MUSEUM OF DIGITAL OLYMPICS, AND AN EXHIBITION CENTRE FOR MANUFACTURERS OF DIGITAL PRODUCTS.
디지털 베이징은 추후에 디지털 올림픽 가상 뮤지엄과 디지털 제품 생산업체를 위한 전시시설로 기능할 것이다.

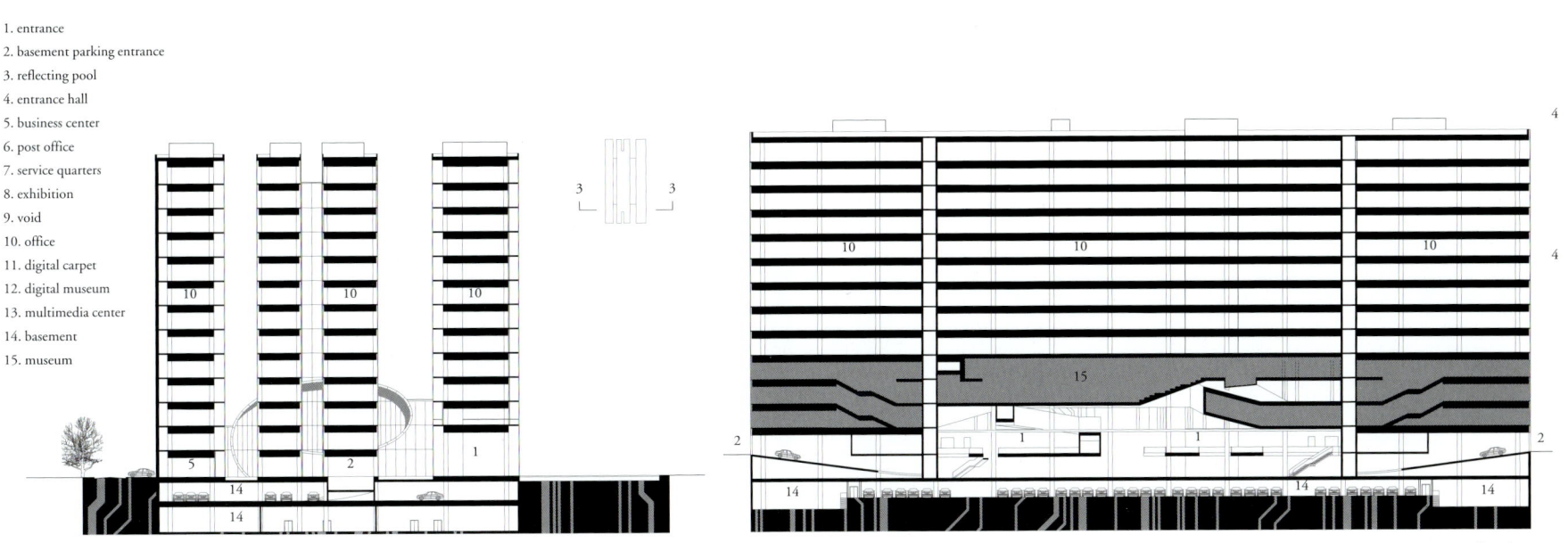

1. entrance
2. basement parking entrance
3. reflecting pool
4. entrance hall
5. business center
6. post office
7. service quarters
8. exhibition
9. void
10. office
11. digital carpet
12. digital museum
13. multimedia center
14. basement
15. museum

Section

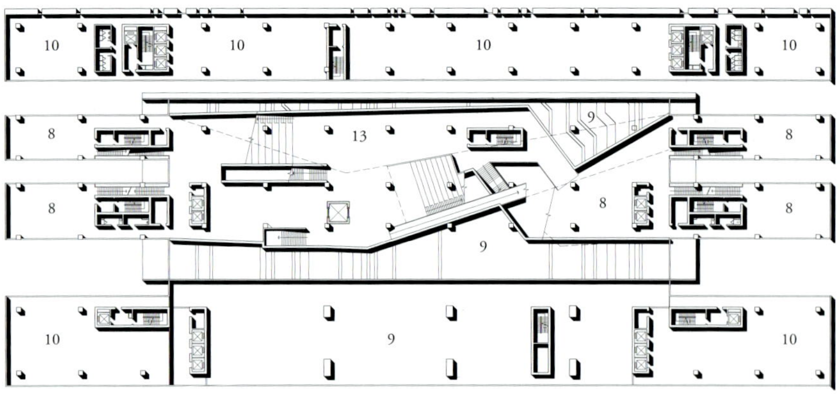

3F plan

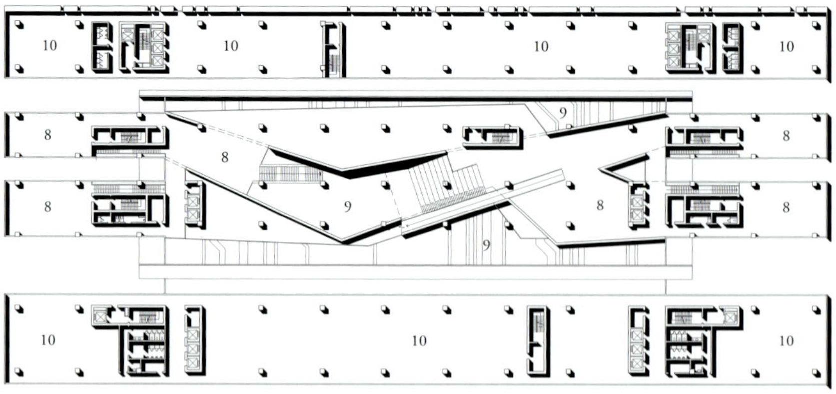

4F plan

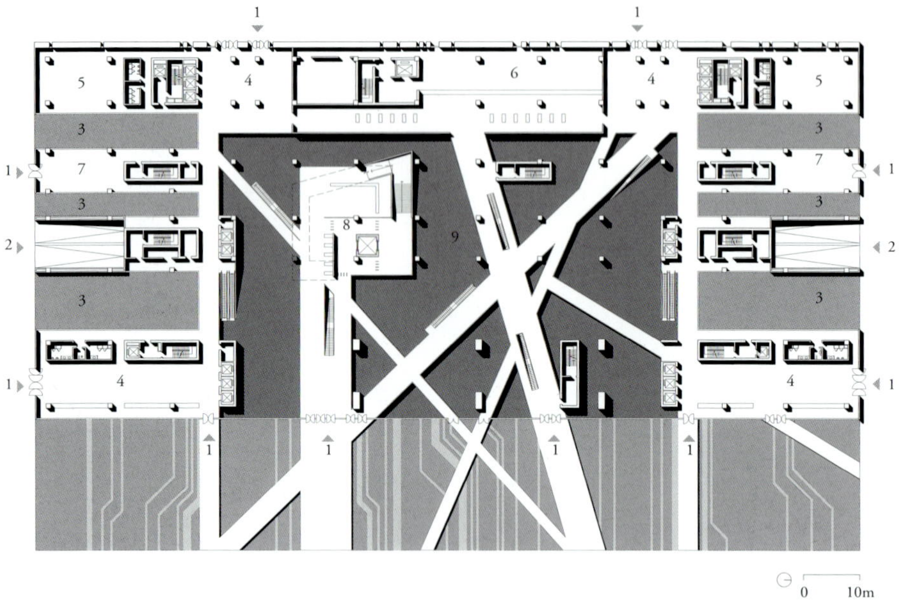

1F plan

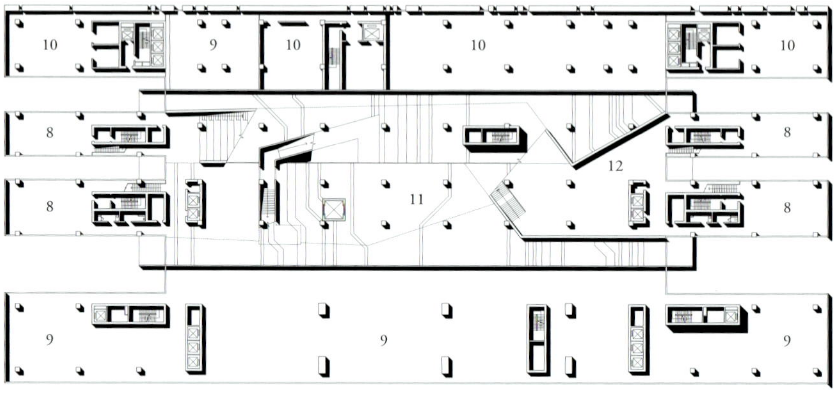

2F plan

1. entrance
2. basement parking entrance
3. reflecting pool
4. entrance hall
5. business center
6. post office
7. service quarters
8. exhibition
9. void
10. office
11. digital carpet
12. digital museum
13. multimedia center
14. basement
15. museum

0 10m

36

HERE WE ARE SEEKING A SPECIFIC FORM, TRYING TO REVEAL AN ENLARGED MICRO WORLD, SUGGESTIVE OF THE MICROCHIPS THAT ARE ABUNDANT BUT IGNORED IN OUR DAILY LIVES.

우리는 구체적인 형태를 원했으며, 우리 삶을 가득 메우고 있지만 쉽게 간과되는 마이크로칩을 연상시키듯 아주 작은 세계를 확대해서 보여주고 싶었다.

GUGGENHEIM ART PAVILION

구겐하임 아트 파빌리온

Saadiyat Island Cultural District, Abu Dhabi

Architects: Zhu Pei, Wu Tong
Associates in charge: Mark Broom, Zeng Xiaoming
Project team: Colop-Morales Frisly, Fan He, Chao Yang, Dong Xue
Structure: Exoskeleton structure
Floor: 2F
Total floor area: 3,500m²
Structural consultant: Rory McGowan (Arup)
Design period: 2006–2007
Construction period: 2009–2011
Client: Guggenheim Foundation

아트 파빌리온은 새로운 문화지구의 도시 조직과 비엔날레 공원 중심에 자리한 운하를 역동적으로 연결하며 대지의 잠재력을 최대한 활용하고 있다. 전시 공간을 비엔날레 공원 위로 올림으로써 연속적인 공공영역이 지상층에 생성된다. 거리, 공원 또는 운하를 통해 접근할 수 있는 이 공간은 전시실로 이어지는 출입구 역할뿐 아니라 운하를 바라보며 커피를 마시거나 소규모 공공전시 또는 공연을 위한 장소로도 사용될 수 있으며, 거리에서 운하로 통하는 환상 교차로 기능도 한다. 전시 공간을 둘러싼 벽은 공원 쪽으로 뻗어 나와 공원과 파빌리온을 즐기는 사람들에게 그늘을 제공하며 거리에서 공원을 감상하도록 돕는다.

건물의 역동적이고 조형적인 언어는 주변의 상업시설이나 주거시설과 차별된 모습으로 문화적 랜드마크로서의 정체성을 강조하며, 도시 조직과 운하를 연결하는 교량으로서의 기능을 형태에 반영하고 있다. 거리의 모습과 스케일을 반영한 북측의 주 출입구는 파빌리온이 있는 비엔날레 공원과 도시의 관계를 강조하듯 자리 잡고 있다. 남쪽으로 좁아지는 파빌리온은 운하를 향해 뻗으며 도시와 물길의 활력 넘치는 관계를 묘사하고 있으며, 하나의 거대한 유리창을 통해 운하와 공원을 감상하게 한다.

극적인 오픈 스페이스로 이루어진 주 전시실은 건물을 수직으로 관통하는 2개의 기울어진 코어를 담고 있다. 공공공간에서 조형적인 열린 계단을 통해 올라오는 주 전시실은 벽과 천장의 주름으로 독특하고 강렬한 조형적 정체성을 유지하는 동시에 유동적으로 사용할 수 있다. 창문은 가장 작게 해 실내로 유입되는 햇볕의 양을 줄여 빛을 조절하고 채광창과 비엔날레 공원을 내려다보는 남측의 큰 창으로 자연 채광과 통풍을 유도했다. 야외 이벤트가 열리는 작은 다목적 중층을 통해 문화지구와 바다를 내다볼 수 있는, 가볍게 출렁이는 지붕으로 연결된다. 관람객의 동선은 주 전시 공간에서 램프를 통해 지상 레벨의 카페로 내려오는 자연스러운 흐름을 가지게 된다.

사무실, 스튜디오 겸 강의실, 상점, 기타 서비스 시설이 있는 파빌리온의 지하층은 파빌리온 아래 조경에 일부 묻혀 있다. 빛을 유입하고 직원 전용 진입로로 기능하는 안뜰은 거리와 운하 양쪽에서 접근할 수 있다. 상점 공간의 지붕은 외팔보처럼 뻗은 파빌리온 아래 테라스를 형성하고 있으며, 운하를 향해 건물의 지상층을 랜드스케이프로 연장시켜 파빌리온과 주변 간의 연계를 강화시킨다.

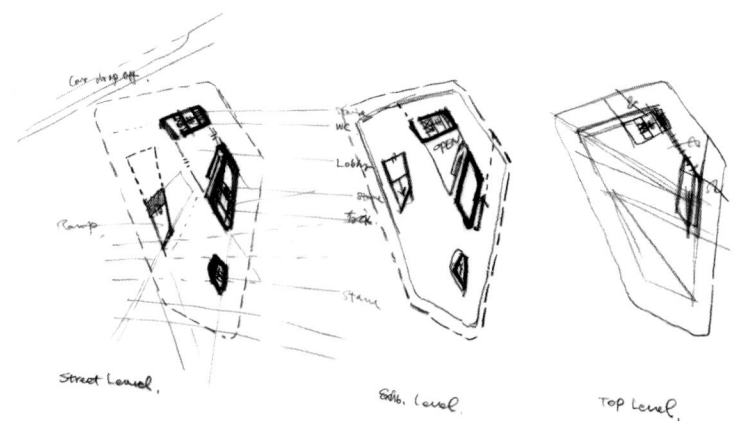

The design of the Art Pavilion aims to maximize the potential of its site by creating a dynamic connection between the urban fabric of the new Cultural District and the canal located at the centre of the Biennale Park. By raising the main exhibition space above the Biennale Park, a continuous public realm is created at ground level. Accessible from the street, park, and canal side, this space is more than just a gateway to the exhibition hall; it is a place for curious passers-by to investigate, a place to enjoy a coffee overlooking the canal, a location for small public exhibitions or performances, or a round-about route from the street to the canal side. The walls, rising to enclose the exhibition space above, slope steeply outward, creating external shaded areas for park and pavilion visitors to enjoy, and for maximizing views of the park from the street.

While the dynamic, sculptural language of the building highlights its identity as a cultural landmark, distinguishing it from the adjacent commercial and residential buildings, the form of the pavilion reflects its function as a bridge between this urban fabric and the canal. To the North, it responds to the geometry and scale of the street. The main entrance is located here, emphasizing the connection of the Biennale Park, of which the pavilion is part, to the wider urban district. To the South, the body of the pavilion tapers, as it appears to stretch out towards the canal, articulating the conception of a vibrant relationship between city and waterway. The building terminates in a single glazed opening, concentrating views across the canal and park.

The main exhibition hall is a spectacular open space, interrupted only by the two angled cores that rise through the building. Accessed by an open, sculptural stair leading up from the public area below, it provides flexibility of use while maintaining a strong and unique formal identity through the folding of walls and roof. Glazing is kept to a minimum in order to reduce solar gain and control light levels, with natural light and ventilation provided by clerestory windows, and the single large window to the south offering views across the Biennale Park. A smaller, multipurpose mezzanine floor leads onto the gently undulating roof, an amphitheatre providing space for outdoor events, and offers views across the cultural quarter and to the sea. A ramp brings visitors back down from the main exhibition space to ground level by the cafe, ensuring a smooth flow of visitors through the building.

Below the main body of the pavilion a basement level that houses offices, flexible studio spaces/classrooms, retail space and services, is partially sunk into the landscape below the pavilion. A courtyard provides light and independent access for staff, and is accessible from both the street level and canal side. The roof of the retail units form a terrace shaded by the cantilevered pavilion. Overlooking the canal, this allows the ground floor of the building to extend out into the landscape, reinforcing the connection between the Pavilion and its surroundings.

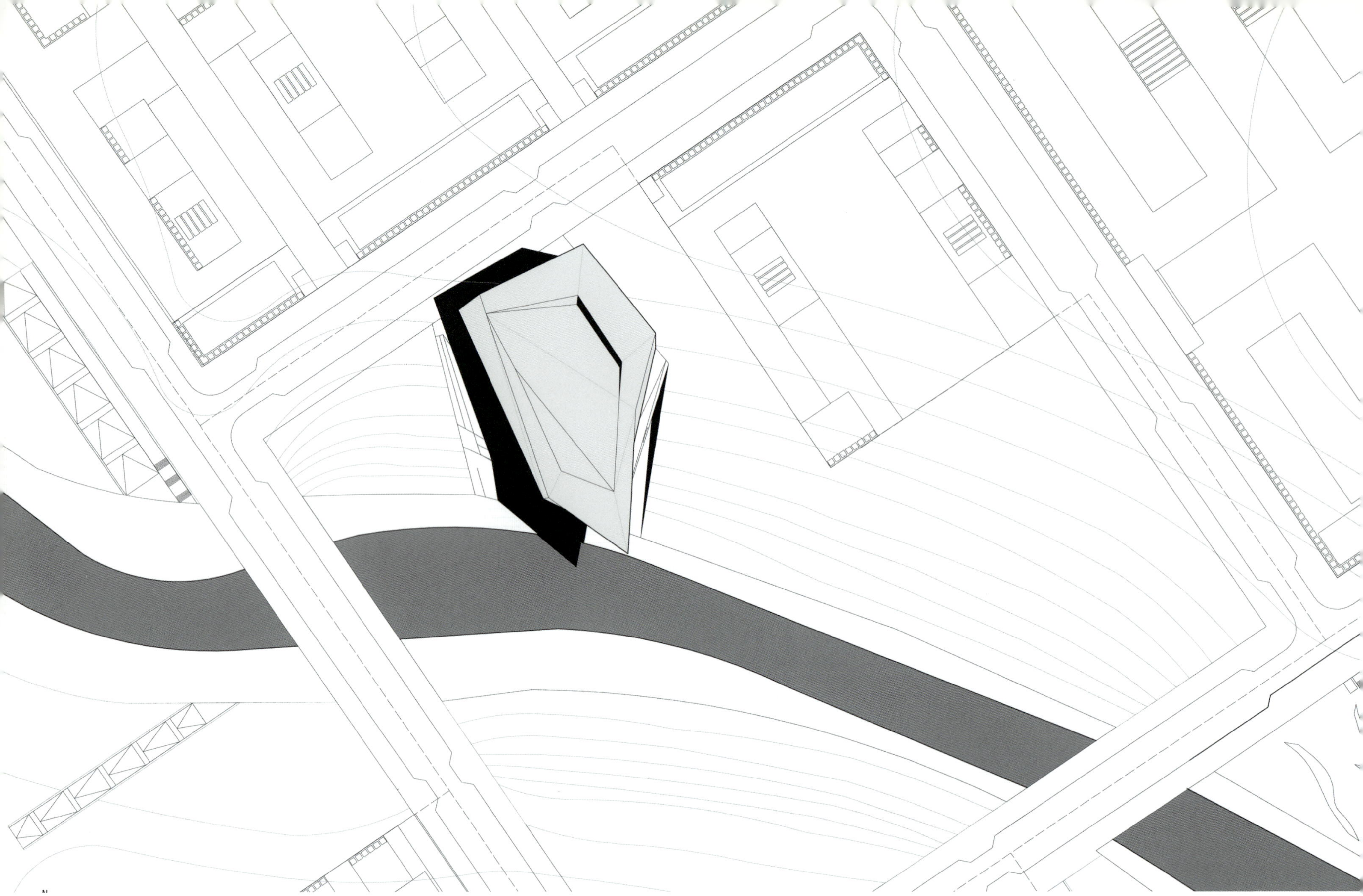

Site plan

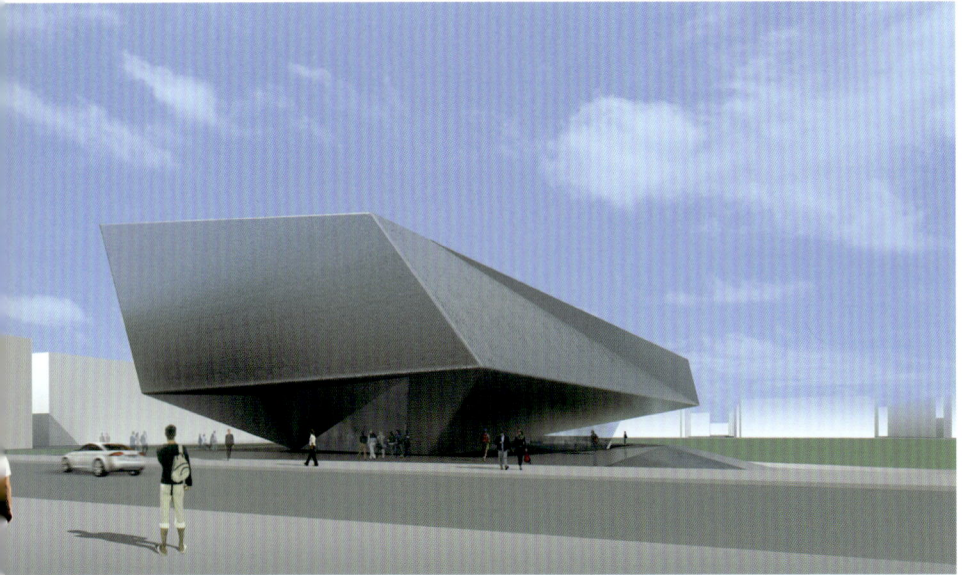

1. exhibition
2. office
3. retail
4. lobby
5. cafe
6. toilet
7. amphitheatre
8. terrace
9. studio
10. void

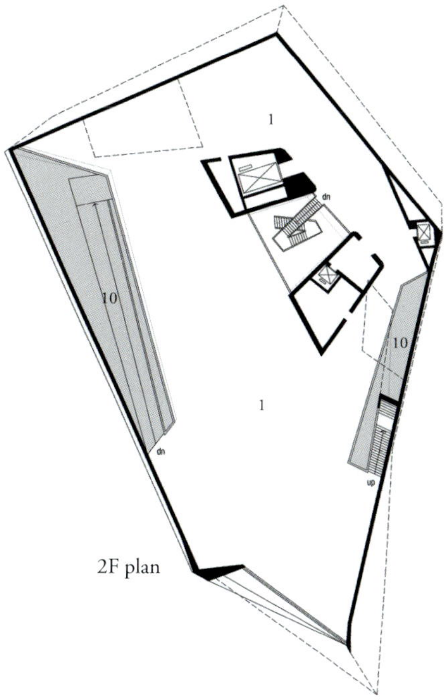

2F plan

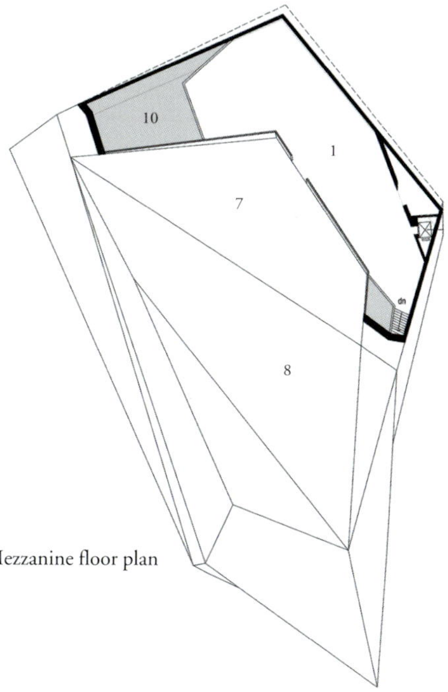

Mezzanine floor plan

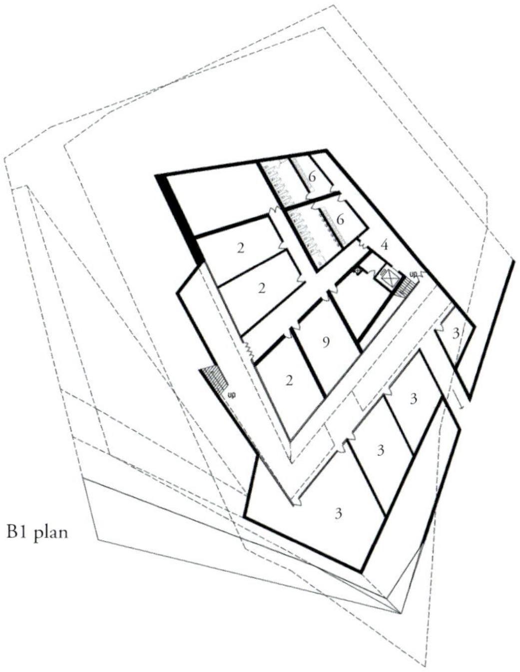

B1 plan

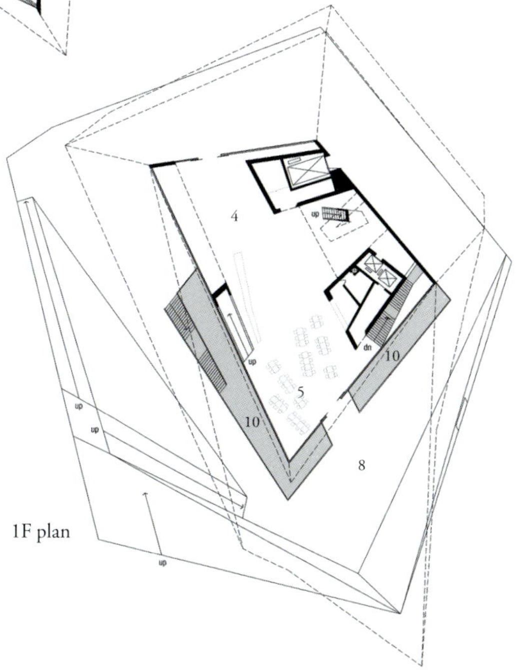

1F plan

1. exhibition
2. office
3. retail
4. lobby
5. cafe
6. toilet
7. amphitheatre
8. terrace
9. studio
10. void

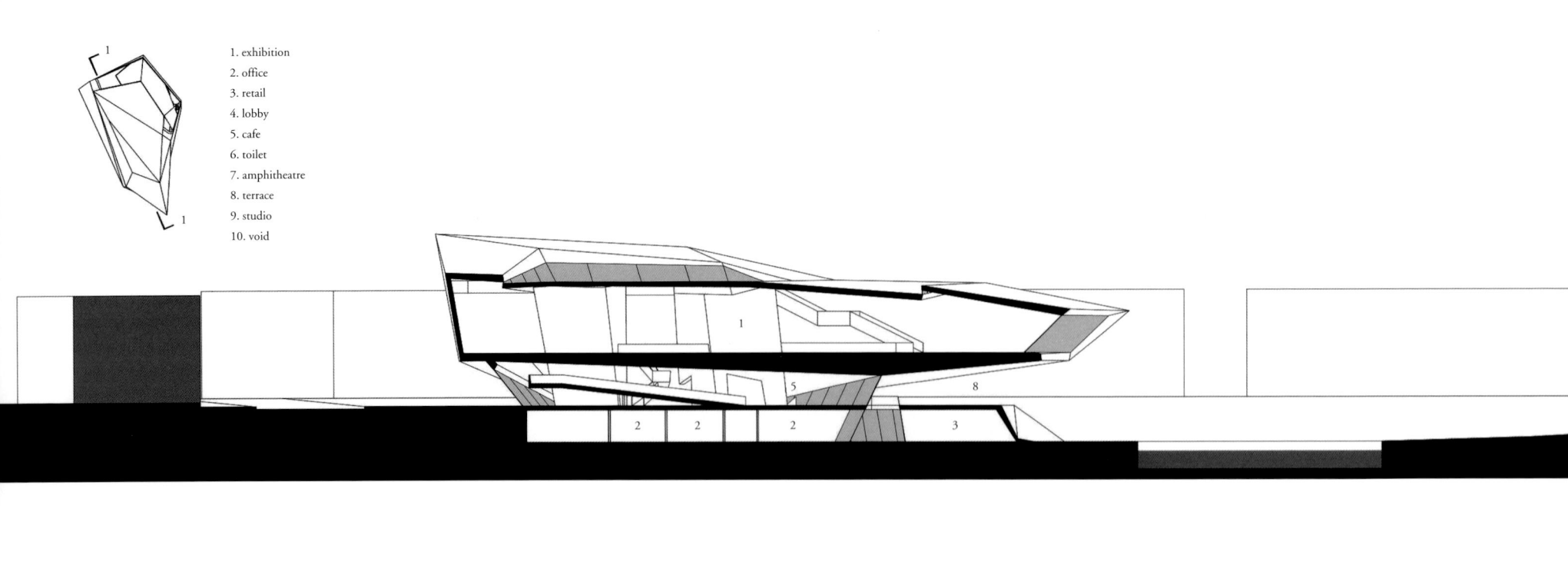
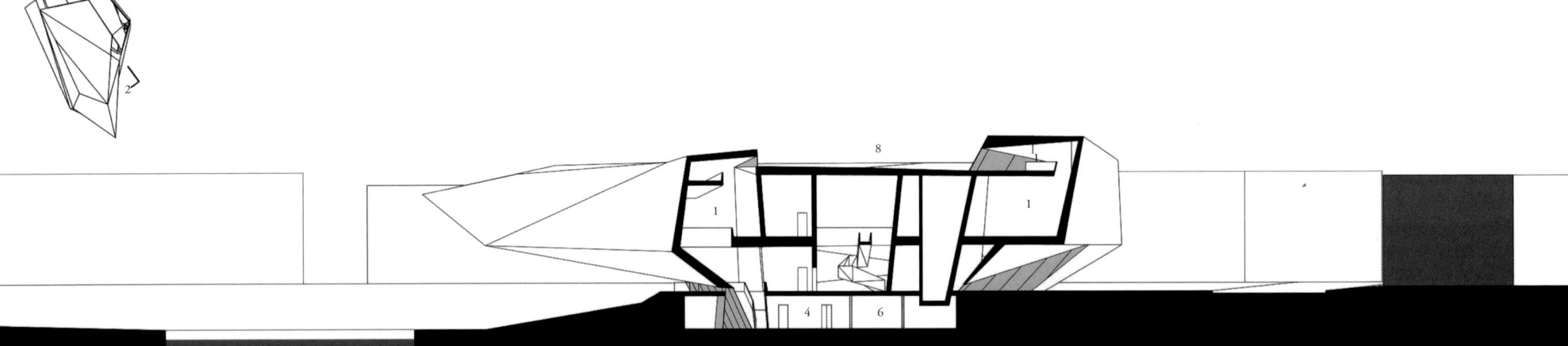
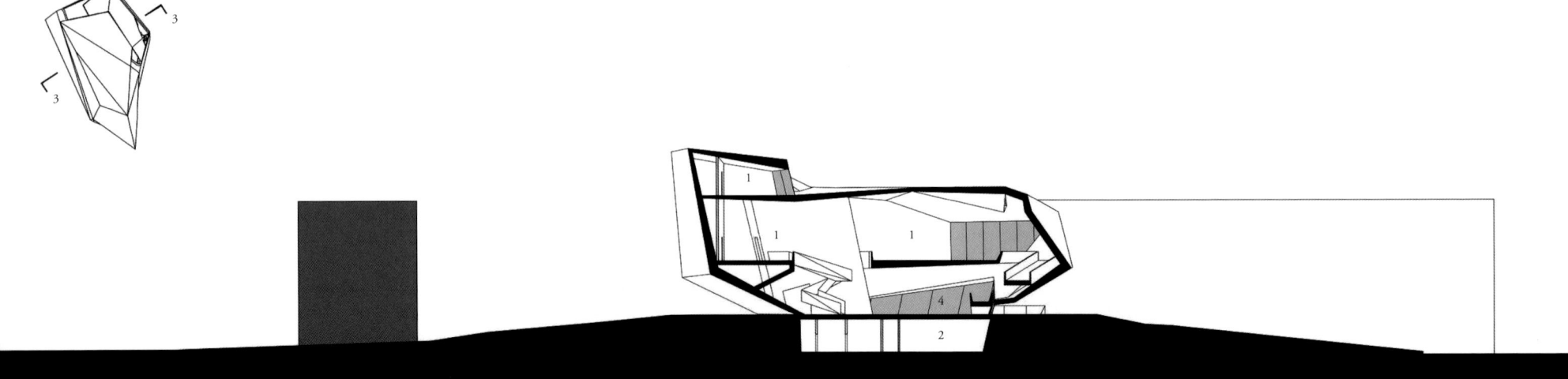

Section

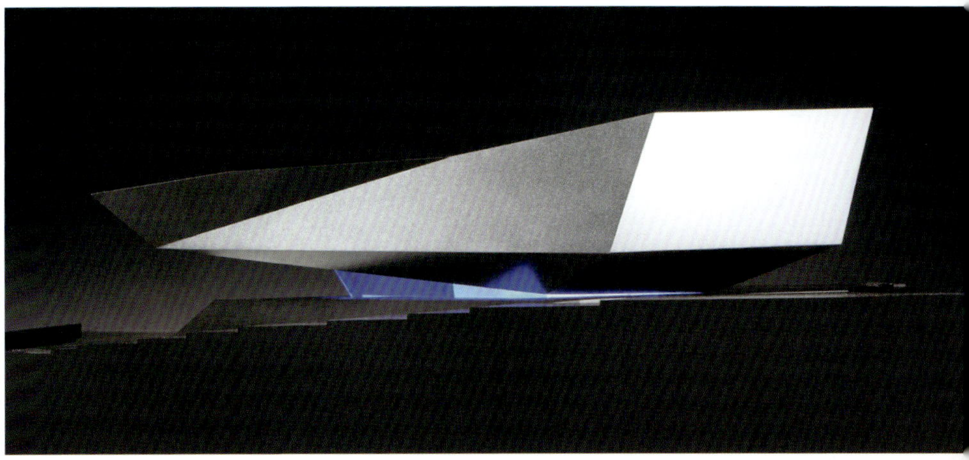

Elevation

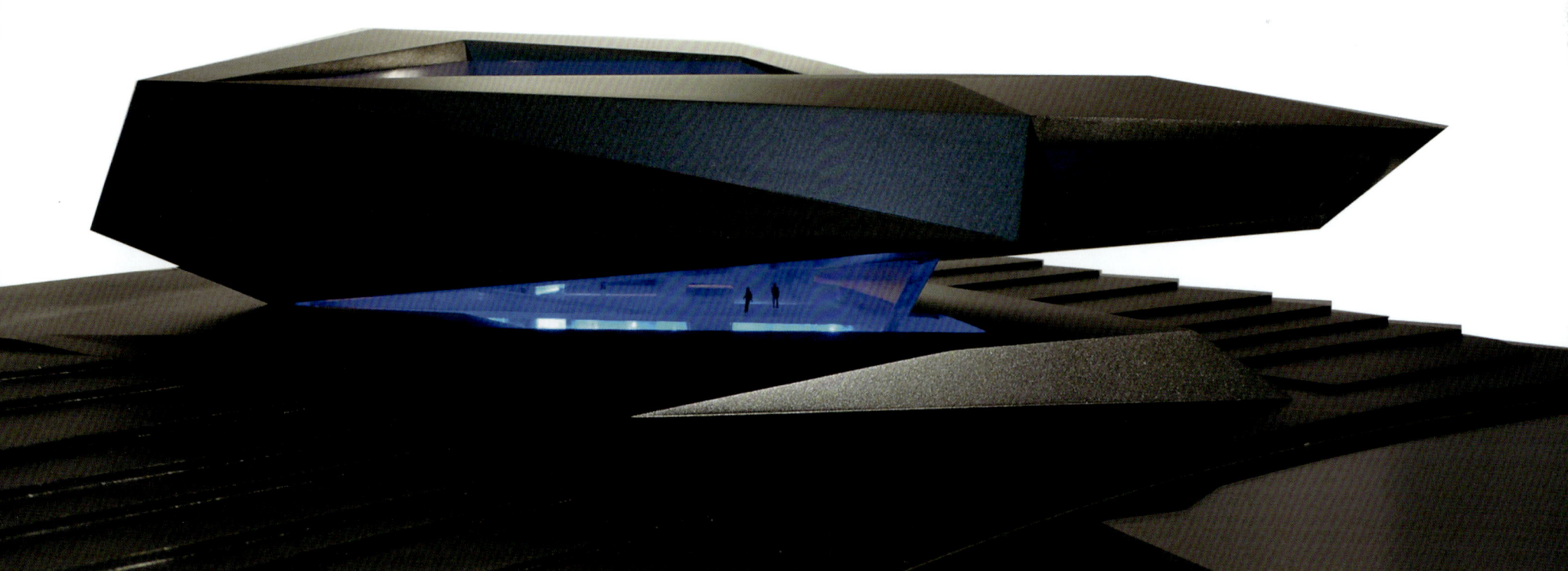

ART MUSEUM OF YUE MINJUN

위에민쥔 미술관
Beijing Olympic Park, Beijing, China

Architects: Zhu Pei, Wu Tong
Project team: Zeng Xiaoming, He Fan, Jiao Chongxia, Li Yongquan, Jiao Chongxia, Fan Xuelan
Structure: Reinforced concrete and Steel frame
Floor: 3F
Site area: 1,000m²
Structural consultant: Rory McGowan (Arup)
Design period: 2006-2007
Construction period: 2009-2011
Photo by: Fang Zhenning

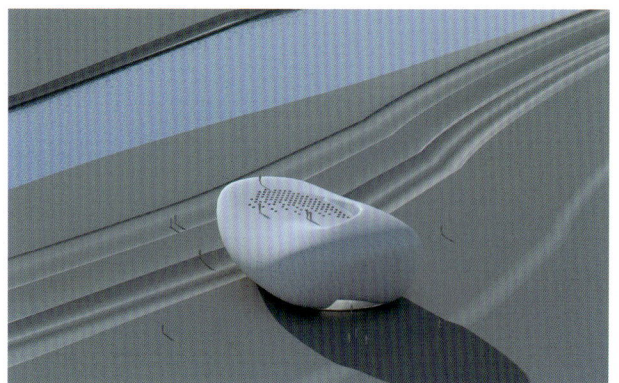

개울과 안개가 감싸는 칭청산의 아름다운 풍경 속에 매우 사적인 예술적 태도와 현대적 특징을 가진 사설 아트 갤러리들이 있다. 과거를 돌아보고 미래를 꿈꾸다 보면 우리는 칭청산에서 만나고 부딪치는 두 가지 상반된 개념이 존재하는 것을 알 수 있다. 그와 같이 흥미롭지만 조화롭지 않은 만남은 어떤 건축 언어로 표현할 수 있을까?

가까이에 자리한 두지앙 옌은 2000년이 넘는 역사 속에서 '자연과 어울리고 인간과 자연이 하나 되는' 것을 지켜봤으며 이 지역 특유의 자연·문화적 환경을 장려해왔다. 우리는 이 프로젝트를 통해 자연을 물려받은 유산으로 생각하고 현실과 상상, 자연과 기술, 전통과 미래 같은 상반된 개념들을 연결시키려 시도했다.

마치 강에서 주운 자갈처럼 매끈하고 유기적인 형태의 미술관 옆으로 스멍강이 흐르고 있으며, 가벼운 금속 코팅으로 처리된 자연적인 형태는 주변 풍경을 살며시 반사하며 자연 속으로 녹아든다. 이는 마치 미래의 비행물체가 지면 위에 떠 있는 듯한 모습을 연상시키기도 한다. 미술관은 예술가의 개인적인 태도를 직접적으로 반영하고 있으며, 미래의 언어로 과거와의 비밀스런 소통을 시작하며 자연과 하나됨을 시도한다.

On one hand, there are the fantastic landscapes of Qingcheng Mountain, with continuous brooks and wreathed mist… On the other hand, there are private art galleries, with extremely personal art attitudes and contemporary features…

When tracing back to ancient times and dreaming of the future, we find two contrary propositions that meet and collide at Qingcheng Mountain. What type of architecture language can be used to describe such incompatible and interesting encounters?

Not far away, Dujiang Weir, with its more than 2,000 years of history, has been witnessing the idea of "complying with nature and combining human life and nature," and also fostering a special natural and cultural environment for this area. In this project, we try to talk with nature in an inheritance and try to create a medium able to overlap reality and imagination, nature and technology, tradition and future, elements which at first seem to be in opposition.

The art museum, located by the Shimeng River, is designed into an organic form full of smoothness and diversity, which is like a cobble taken from the river. A light metal coating is applied to this ancient natural form, mildly reflecting the surrounding scenery as it melts into nature, making the building seem suspended over the ground, like a flying body from the future. We present an art museum that features the direct and affirmed personal attitudes of the artists, that participates in nature in a "lost" way and starts a secret dialogue with ancient times in a language of the future.

A LIGHT METAL COATING IS APPLIED TO THIS ANCIENT NATURAL FORM, MILDLY REFLECTING THE SURROUNDING SCENERY AS IT MELTS INTO NATURE.

가벼운 금속 코팅으로 처리된 자연적인 형태는 주변 풍경을 살며시 반사하며 자연 속으로 녹아든다.

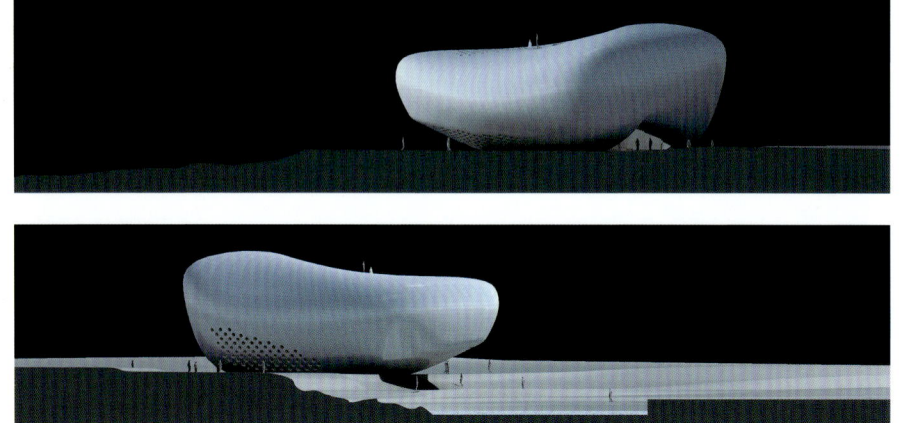

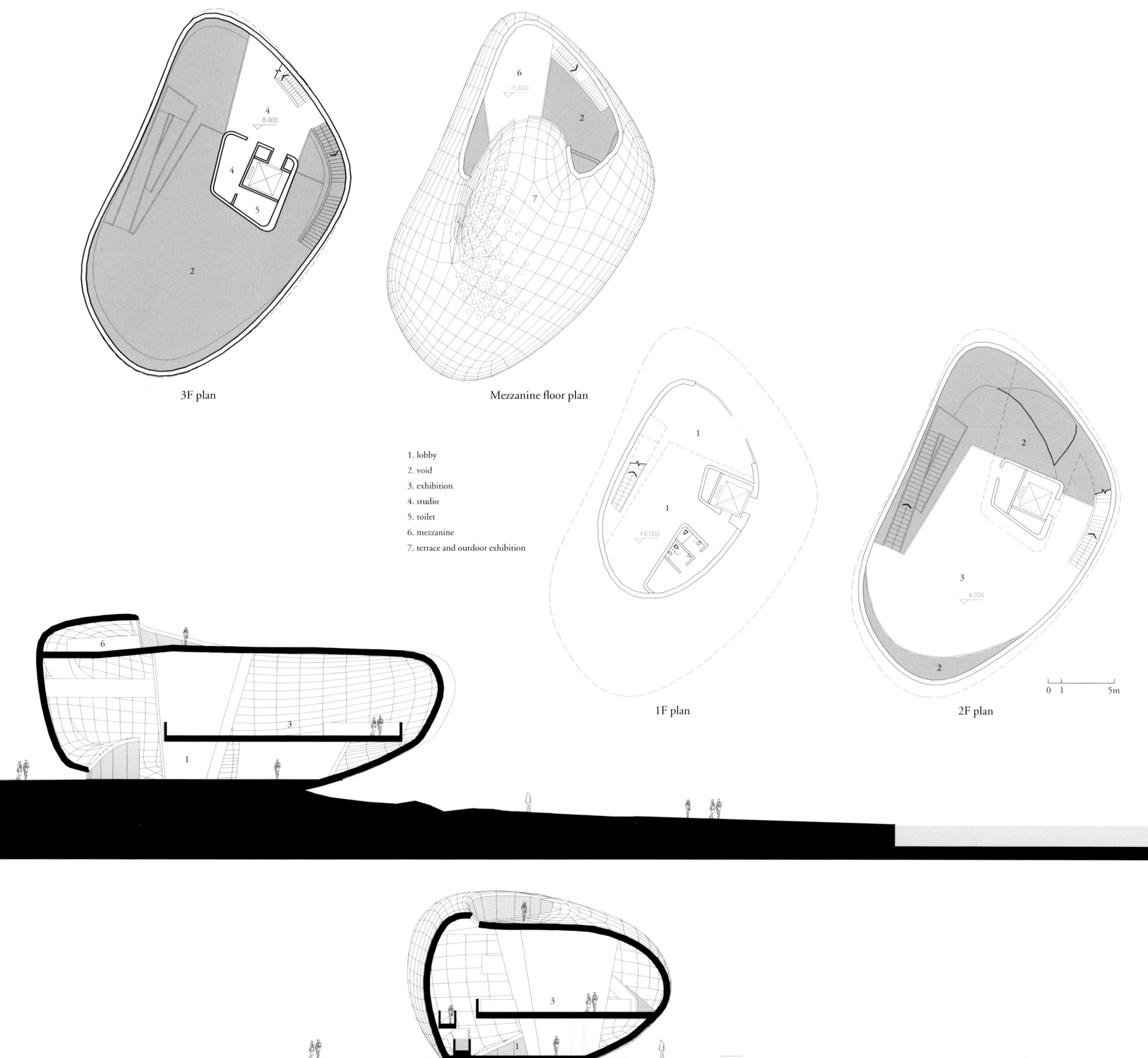

3F plan Mezzanine floor plan

1. lobby
2. void
3. exhibition
4. studio
5. toilet
6. mezzanine
7. terrace and outdoor exhibition

1F plan 2F plan

Section

PUBLISHING HOUSE

퍼블리싱 하우스

North Ring, Beijing, China

Architects: Zhu Pei, Mark Broom, Lu Wei
Project team: Mark Broom, Frisly Colop-Morales, Fan He, Chao Yang, Jiao Chongxia, Dai Lili, Xi Weidong, Li Shaohua, Lu Wei
Structure: Reinforced concrete frame
Floor: 14F
Site area: 11,000m²
Total building area: 9,900m²
Design period: 2007
Construction period: 2007-2008
Client: Beijing Publishing Group

퍼블리싱 하우스는 북경출판공사가 사용 중인 12층짜리 사무실 건물을 출판 산업과 창의력을 장려하는 새로운 센터로 탈바꿈시키는 프로젝트다. 출판에 대한 대중의 관심을 높이려는 목적의 이 건물은 자금성과 올림픽 공원을 남북으로 관통하는 축과 북경의 제3순환도로 북쪽이 교차하는 주요 지점의 코너에 있다.

전 세계적으로 큰 영향을 주고 있는 기술과 텔레비전, 인터넷의 발달로 인쇄 매체의 역할이 점차 줄면서 현대 인쇄 산업은 빠르게 변화하고 있다. 중국의 경제 개혁에 따른 사회적 변화도 출판업계에 영향을 주고 있다. 경제가 활발해지고 중산층이 늘어나면서 여가·부·교육 수준이 증가하고, 이는 모든 형태의 매체에 더 큰 가치를 부여하며 도시 내 가정이나 거리 모두에서 보다 유비쿼터스적인 존재성을 주었다. 출판업계가 현시점에서 살아남기 위해서는 그와 같은 패러다임의 이동에 빠르게 적응해야 하며, 도시의 실제 공간과 가상공간에서 일어나는 상호작용을 출판 산업의 창의적인 원동력으로 활용할 수 있는 새로운 방향을 찾아야 한다.

그러한 생각으로 시작한 프로젝트는 출판을 주제로 한 '마이크로시티'라는 하나의 건물 안에 주변 도시 조직의 소세계를 만들었다. 기존의 일괄적으로 구획된 사무실 건물은 사무·교육·상업 및 레저 공간들이 어우러진, 다양한 성격이 혼재된 건물로 재탄생한다. 일반인에게 개방되지 않았던 기존 건물에 공공공간과 공용공간을 마련했으며, 수직으로 관통하는 커뮤니케이션 통로로 층간에 열린 관계를 만들었고, 외부 공간을 통해 건물 사용자와 도시의 연결을 도모했다. 또한 다양한 사이즈의 외팔보를 더해 기존 층을 연장시켰는데 이들은 북측과 서측에 집중되어 있으며, 기존 철근 콘크리트 구조를 최소한으로 변경했다. 이들은 여러 층을 연결시키며 외부에 테라스 공간을 만들었다. 건물을 시각적 매체로 해석한 전략은 건물의 화려한 형태에서도 읽을 수 있다. 기존 건물의 복도를 막고 쌓여 있던 책들에서 영감을 얻은 형태로 실내와 입면에 패턴을 만들고 있다. 건물 안에서 일어나는 여러 활동과 공간들이 만나며 발생하는 상호작용과 시각적·물리적·프로그램적인 도시와의 연결을 통해 건물 안의 삶뿐 아니라 주변 도시 조직의 삶까지도 활력을 얻길 바란다.

Publishing House is a project that aims to transform an existing 12-storey office building occupied by the Beijing Publishing Corporation into a centre for promoting creativity within the publishing industry. It also aims to increase public interest in publishing. The site has a prominent corner location on Beijing's busy northern Third Ring Road, on the north-south axis running between the Forbidden City and the Olympic Park.

The contemporary industry is changing rapidly. As is the case all over the world, technological advances have greatly affected the media, with the rise of television and the Internet diminishing the role of the printed media. Societal transformations related to China's economic reforms are also affecting publishing. As the economy booms and the middle classes expand, increased leisure time, affluence, and education means an increased value placed upon the media in all its forms. The media, as a result, has assumed a more ubiquitous presence throughout the city, both in people's homes and on the streets. The publishing industry must adapt swiftly to this paradigm shift in order to survive, and it is here, in the interactions taking place in the real and virtual spaces of the city, that the creative impetus for new directions in publishing can be found.

Working from this premise, the project seeks to create a microcosm of the surrounding urban fabric within one building - a MicroCity focused on publishing. The existing homogenous, compartmentalized office building will be transformed into a heterogeneous mix of linked spaces for work, learning, retail, and leisure. Public and communal space is introduced into a previously closed building, vertical paths of communication between floors are opened up, and outdoor spaces connecting the occupants with the city are created. In order to achieve this, a series of cantilevers of varying sizes are constructed, extending the existing floors. Concentrated on the north and west aspects, these allow links to be formed between floors, with only minimal modification of the existing reinforced concrete structure. They also create external terraces. This strategy of building as visual media can also be seen in the dramatic form of the building, which echoes that of the stacks of books found clogging up the corridors in the existing building, and in the proposed use of patterning in the interiors or the facade. It is hoped that the opportunities for interaction found in the intersections between the spaces and activities contained within the building, and the connections (at once visual, physical, and programmatic) made with the city beyond will foster a creative and vibrant atmosphere that can stimulate life not only within the building itself, but also in the urban fabric that surrounds it.

Photographs by Fang Zhenning

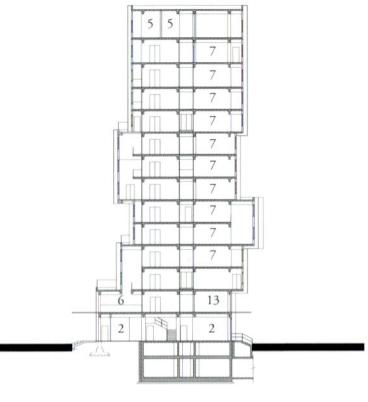
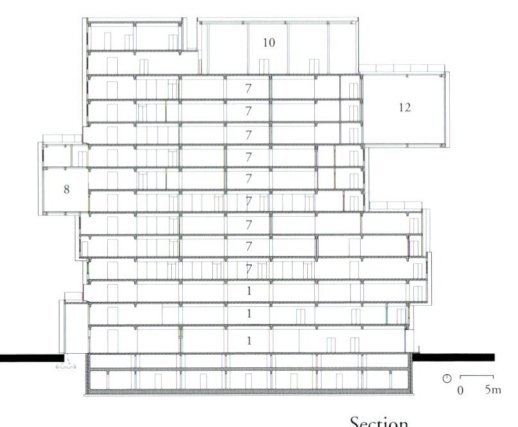

Section

1. book shop
2. lobby
3. cafe
4. bar
5. mechanic
6. restroom
7. office
8. exhibition
9. kitchen
10. chamber
11. terrace
12. multipurpose room
13. operating room

1. book shop
2. lobby
3. cafe
4. bar
5. mechanic
6. restroom
7. office
8. exhibition
9. kitchen
10. chamber
11. terrace
12. multipurpose room
13. operating room

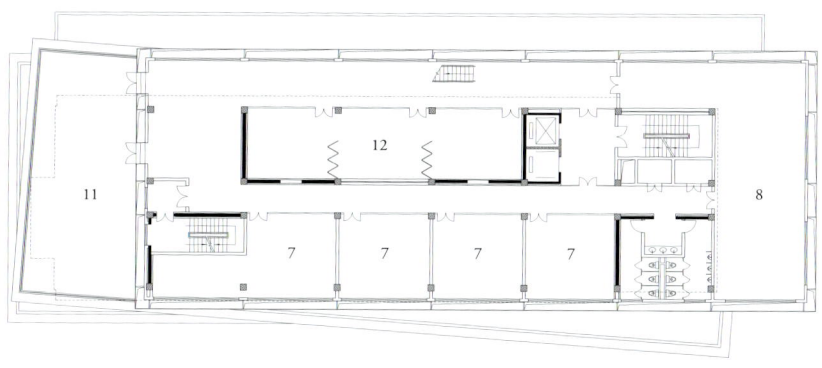

7F plan

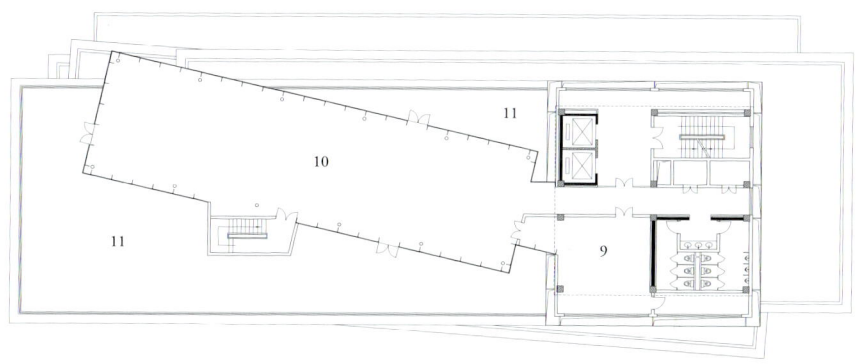

13F plan

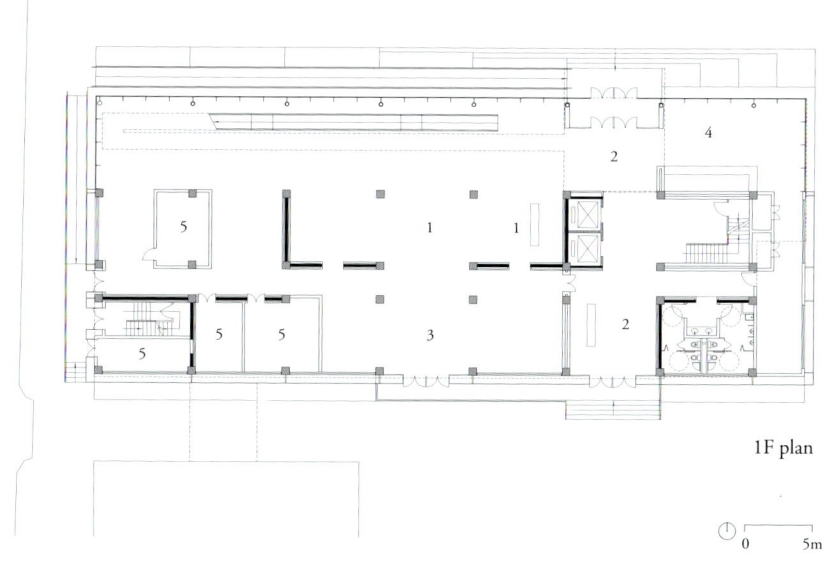

1F plan

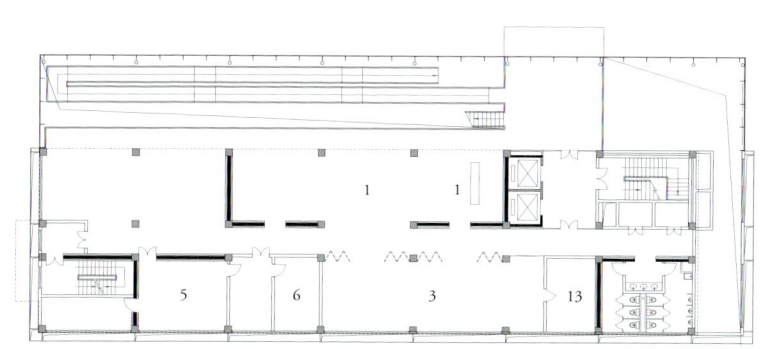

2F plan

0 5m

BLUR HOTEL

블 러 호 텔

Beijing, China

Project designer: Zhu Pei, Wu Tong
Project team: Li Chuen, Zhang Pengpeng, Zhou Lijun, Dai Lili, Wang Min
Program: Business hotel, Cultural facility
Structure: Reinforced concrete frame (existing structure)
Building area: 10,176m^2
Design period: 2004-2005
Construction period: 2005-2006
Engineer: Beijing Zhongjian Hengji Gongcheng Co. Ltd.
Client: China Resource

베이징은 한때 세계에서 가장 잘 보존된 중세 도시였다. 하지만 1949년 이후 옛 도시의 리듬을 무시한 채 대규모 정부시설과 산업시설이 도심에 들어서며 원활하게 흐르던 도시의 위계질서를 흐트러뜨렸다. 그와 같은 개발은 도시의 '종양' 같은 존재로 고대 명 왕조 시대부터 이어져 온 도시를 불균형적이고 혼란스러운 도시로 변화시켰다.

자금성 서문 옆에 자리한 커다란 정부 행정시설 부지에 들어선 블러 호텔은 그런 문제를 치유하기 위한 '도시 침술'적인 실험이다. 시술로 종양을 제거하는 것보다는(다른 말로 또다시 허무는 것보다는) 훨씬 덜 파괴적인 방법으로, 그대로 둔 채 해로운 성질을 중화시키는 방식을 택했다. 프로젝트는 기존 건물과 주변을 조화롭게 재개발할 것을 제안했고, 구식으로 옛것을 모방하는 것이 아닌 주변 지역을 위한 재개발에 대한 좋은 선례가 되길 목표했다.

그에 따른 첫 번째 전략은 공공성격의 프로그램 위주로 일반인에게 개방된 공간을 만들어 건물의 지상층을 여는 것이었다. 두 번째는 지역 특유의 전통 건축 양식인 사합원, 즉 마당 있는 주택과의 조화를 꾀했다. 기존 건물의 콘크리트 슬라브를 잘라 주변 후통의 공간적 구성을 닮은 수직 안뜰이 반복되는 구성을 만들었다. 건물의 실내를 변화시킨 뒤 우리가 취한 마지막 전략은 건물 외부를 반투명 파사드로 둘러싸는 것이었다. 지역 전통을 반영한 이 스킨은 중국의 초롱 이미지를 표현하고 있으며, 외부의 빛을 건물 내부로 유입하거나 내부의 빛을 외부로 발산하며 모든 층을 하나의 발산하는 오브제로 만든다.

Beijing was once one of the best-preserved medieval cities in the world. Since 1949, however, the location of government and industrial premises within the centre of the city has disrupted the once free-flowing and hierarchical city plan with the construction of enclosed, large-scale buildings placed with no regard to the rhythm and consistency of the old city. The development of these "tumors" within the ancient Ming dynasty core has resulted in the creation of a disjointed and incomprehensible city centre.

In response to this problem, Blur Hotel, located on the site of a large government office building beside the Western Gate of the Forbidden City, is an experiment in "urban acupuncture." Rather than operate and remove the tumor (in other words demolish yet again), a far less disruptive and harmful method is to leave it in place and simply neutralize its ill effects. As a refurbishment proposal, the project aims to harmonize the existing building with its surroundings without resorting to backward-looking pastiche, and hopes to provide a beacon for renewal of the surrounding area.

The first strategy employed with this end in mind is to open out the ground floor of the building to create a layer of traversable space, occupied by public-oriented programs. The next approach aims to integrate the building more with the local building typology of the Siheyuan, or courtyard house. By simply carving into the concrete slab floors of the existing building, an arrangement of alternating vertical courtyards is created, replicating the spatial arrangement of the surrounding Hutongs. With the interior of the building transformed, the third and final tactic deals with the exterior of the building, wrapping it in a continuous and semi-transparent facade. Referring to local traditions, this skin is based on the image of a Chinese lantern. The allowance of light into and out of the building on every floor diffuses the building into a single, but permeable, object.

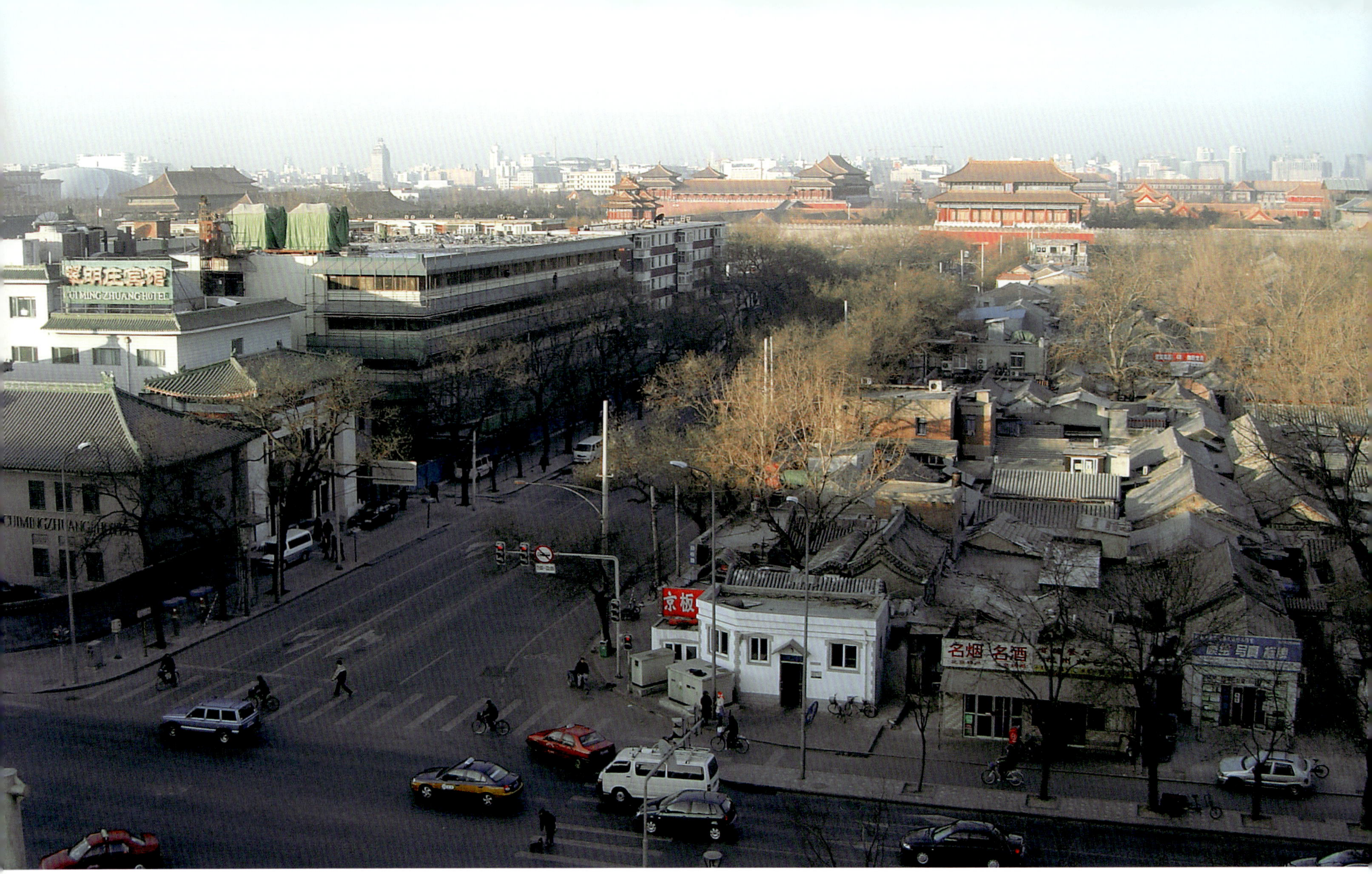
CUI MING ZHUANG HOTEL
名烟 名酒

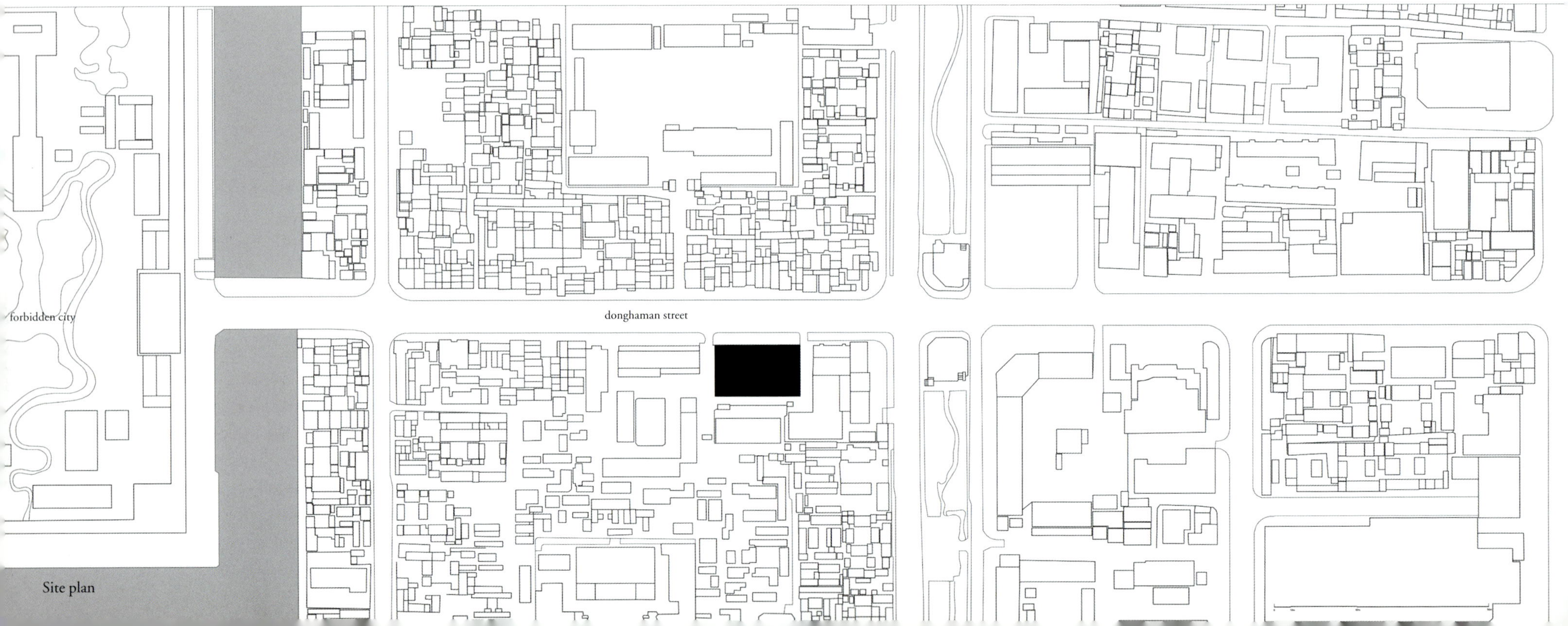
forbidden city
donghaman street

Site plan

REFERRING TO LOCAL TRADITIONS, THIS SKIN IS BASED ON THE IMAGE OF A CHINESE LANTERN. THE ALLOWANCE OF LIGHT INTO AND OUT OF THE BUILDING ON EVERY FLOOR DIFFUSES THE BUILDING INTO A SINGLE, BUT PERMEABLE, OBJECT.

지역 전통을 반영한 이 스킨은 중국의 초롱 이미지를 표현하고 있으며, 외부의 빛을 건물 내부로 유입하거나 내부의 빛을 외부로 발산하며 모든 층을 하나의 발산하는 오브제로 만든다.

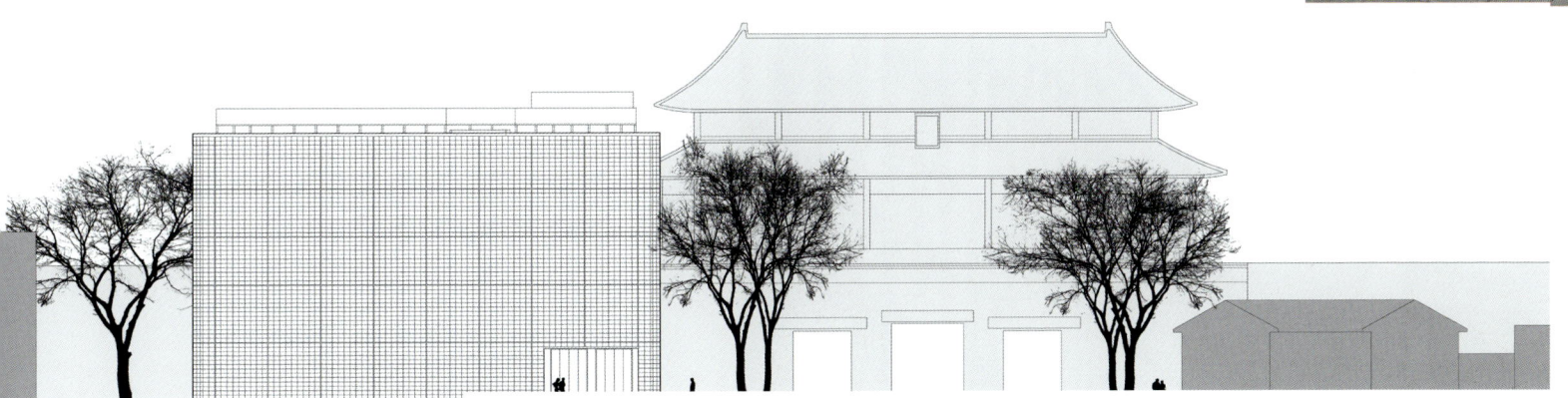

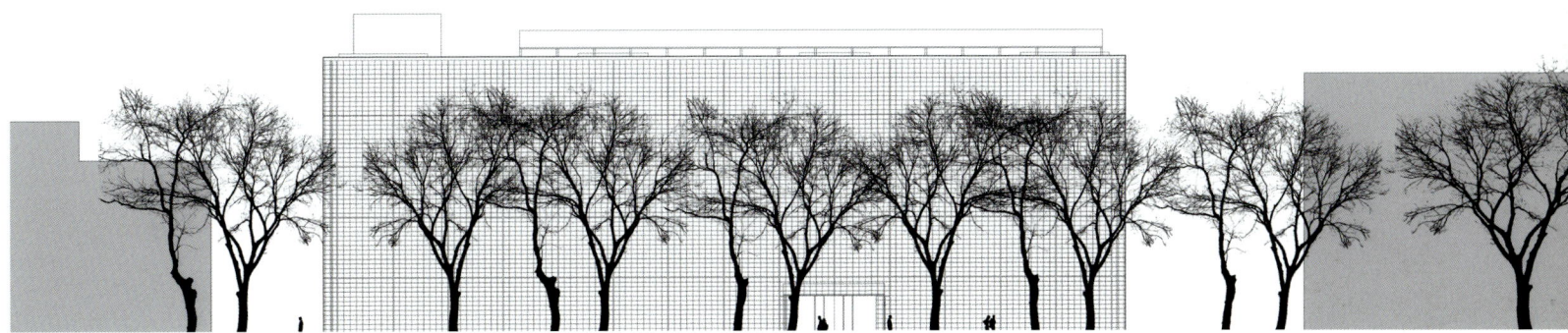

Elevation

As a refurbishment proposal, the project aims to harmonize the existing building with its surroundings without resorting to backward-looking pastiche, and hopes to provide a beacon for renewal of the surrounding area.

기존 건물의 콘크리트 슬라브를 잘라 주변 후통의 공간적 구성을 닮은 수직 안뜰이 반복되는 구성을 만들었다.

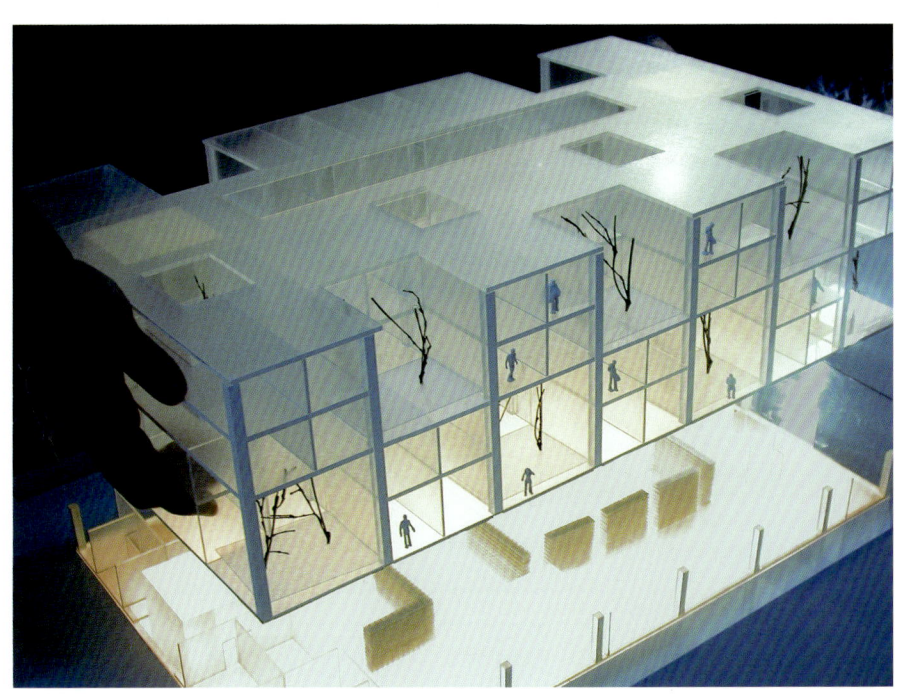
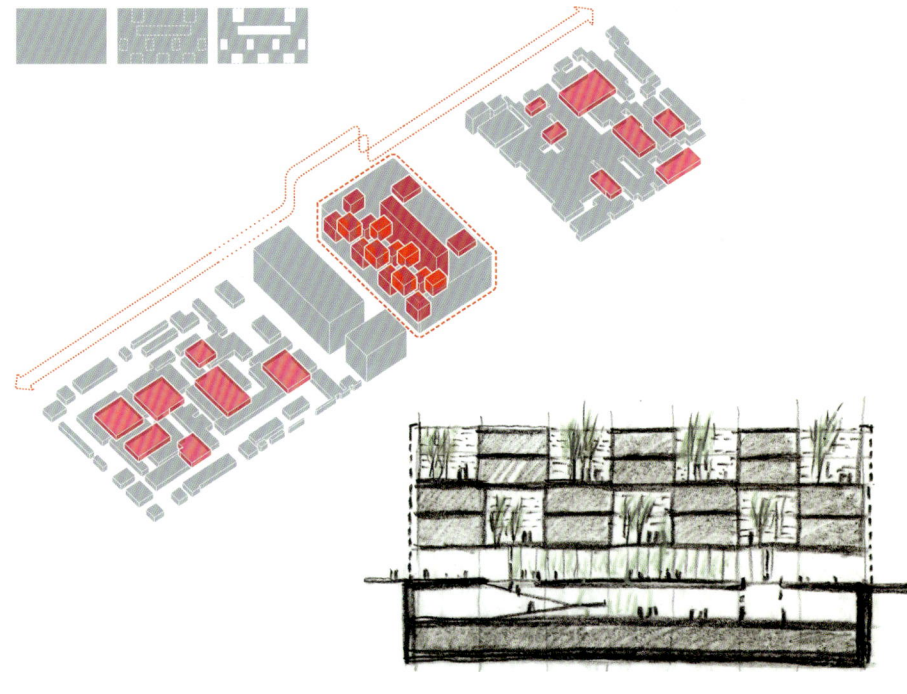

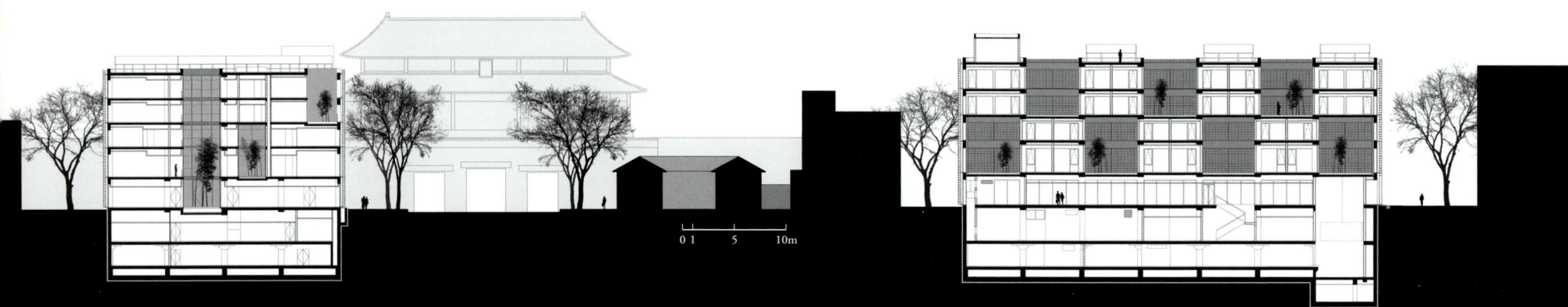

Transverse section　　　　　　　　　　　　　　　Section

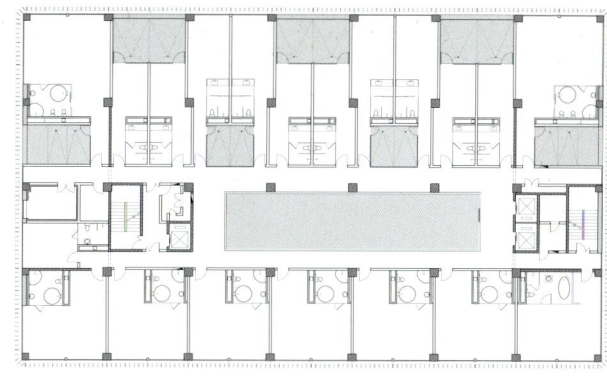

4F plan

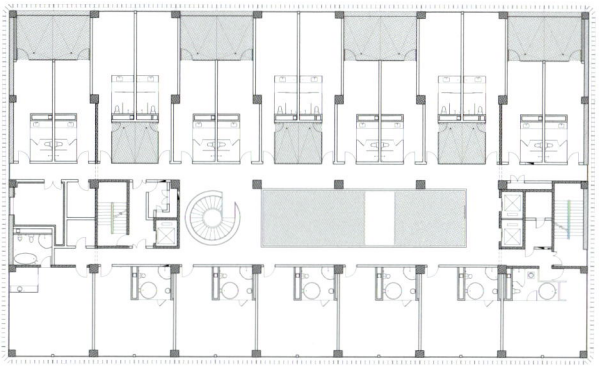

2F plan

CAI GUOQIANG COURTYARD HOUSE RENOVATION

차이 궈치앙 전통 주택 리노베이션

Beijing, China

이 예술가의 집은 '사합원(四合院)'이라는 역사적으로 중요한 중국의 마당 있는 전통 가옥을 복원하고 새로 신관을 더하는 형식으로 진행되었다. 대지는 자금성과 매우 가까운 곳으로 원래 베이징대학 부지였기 때문에 여러 어려움이 더해졌다. 한편 증축을 위해 후통으로 둘러싸인 주변 환경과 내부 마당 구조를 민감하게 고려해야 했다. 우리는 재생에 대한 민감한 입장을 취하고 현대 건축물이 전통적인 구조물과 어떻게 공생할 수 있는지에 대해 고민했다. 여러 중국 도시와 마찬가지로 베이징의 역사적인 중심지 역시 무자비한 '현대화' 속에서 역사적인 전통 구조와의 조화에 어려움을 겪고 있으며, 전통적인 맥락을 벗어난 역사적인 중심의 현대적 사용은 조형적인 조화를 잃었다. 하지만 우리는 현대 구조물이 그 기능은 잃지 않으면서 옛 선조들의 언어를 응용할 수 있다고 믿었다.

지역의 기술자 및 건설업자와 함께 전통적인 재료와 공법을 사용해 원래의 사합원을 예전 상태로 복원했고, 남향 건물에 의도적인 보이드를 두는 2개의 마당을 가진 구조도 복원했다. 오랜 세월에 걸쳐 낡은 내부는 원래의 바닥 타일 및 벽 마감재를 사용해 복원했으며 구조를 노출시켰다. 신관은 더 큰 남쪽 마당에서 전통 가옥을 바라보며 공중에 떠 있는 듯한 오브제로 완성되었다. 자신의 조상과 마주하고 있는 '보이지 않는' 이 건물은 공들여 복원한 가옥과 대립 및 보완 관계에 놓여 있다. 기존의 담과 독립적으로 설계해 옛 구조에 최소한의 영향을 주도록 의도했으며 시각적·물리적으로 가벼운 구조다. 신관은 유리와 스틸 같은 현대적인 재료를 사용했으며, 재료의 반사하는 성질을 이용해 옛 건물에 대한 존경을 표현하고 있다. 또한 기존 건물이 가진 고정된 프로그램과 달리 유동적인 다목적 공간이나 예술가의 작업실 같은 현대적인 프로그램들이 더해졌다. 가볍고 눈에 띄지 않는 신관은 무겁고 위엄 있는 옛 건물과 대조를 이루지만 그들은 형태, 스케일, 기능적인 면에서 서로를 보완하고 있다.

This residence for an artist calls for the restoration of a historically significant classical Chinese Siheyuan courtyard house, and a new building addition within its compounds. This project was particularly challenging because of its site: situated very close to the Forbidden City and also near the original grounds of Peking University, the addition had to be sensitive to its external surroundings, the Hutong neighborhood, and internal courtyard configuration. The attitude we developed for this project was one of sensitive regeneration, where contemporary and modern buildings can symbiotically co-exist with traditional structures. Like many Chinese cities, Beijing's historical core is succumbing to unrelenting 'modern' development because of the difficulty in engaging traditional forms, steeped in history. Its modern uses are formalistically at odds with traditional contexts. However, modern structures can adopt their predecessors' vocabulary, yet still serve modern functions.

The original Siheyuan was restored to its original condition, using traditional materials, construction technologies, and local skilled craftsmen and builders. The double courtyard configuration was restored into its intended void, which gives presence to the south-facing buildings. Because of the disintegration that has taken place over the years, the interior was renovated with the original types of floor tiles, wall surfaces, and structural exposure. The new building, however, was created as an object that floats in the larger south courtyard to face the traditional structures. In dialogue with its immediate predecessor, the new "invisible" building is both oppositional and complementary to the painstakingly restored houses. It is detailed to be independent from the existing courtyard walls, so as to create as little impact as possible on the old structure; it is visually and physically light. Modern materials such as glass and steel are used throughout the new building, so as to take advantage of its reflective properties and to pay homage to the old. It contains new 'modern' programs, such as a flexible multi-purpose space and an artist studio, versus the fixed traditional programs of the existing structure. Similarly, the light and invisible new building stands in contrast to the heavy and commanding presence of the old, but they complement each other in form, scale, and function.

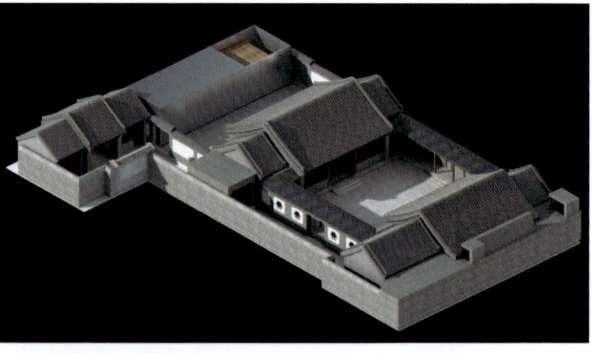
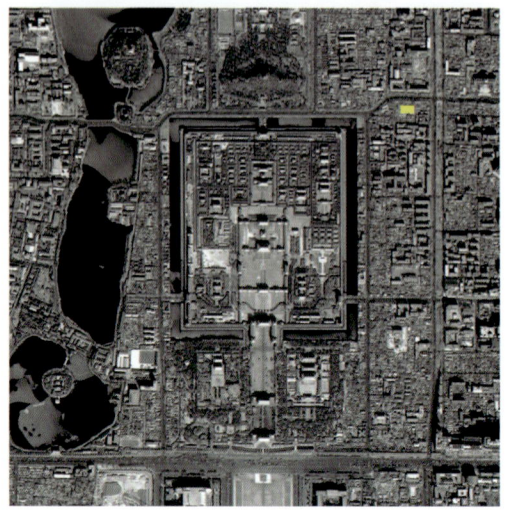

The new building was created as an object that floats in the larger south courtyard to face the traditional structures.
신관은 더 큰 남쪽 마당에서 전통 가옥을 바라보며 공중에 떠 있는 듯한 오브제로 완성되었다.

BEIJING XISI BEI REGENERATION STRATEGY

베이징 시쓰베이 재생전략

Architects: Zhu Pei, Wu Tong
Associates in charge: Mark Broom, Zeng Xiaoming
Design team: Xue Dong, Evelina Sausina, Lu Wei, Xi Weidong, Lucas Ledderose, Sina Keesser, He Fan, Zhao Wei, Li Xiang, Liu Xiaofei

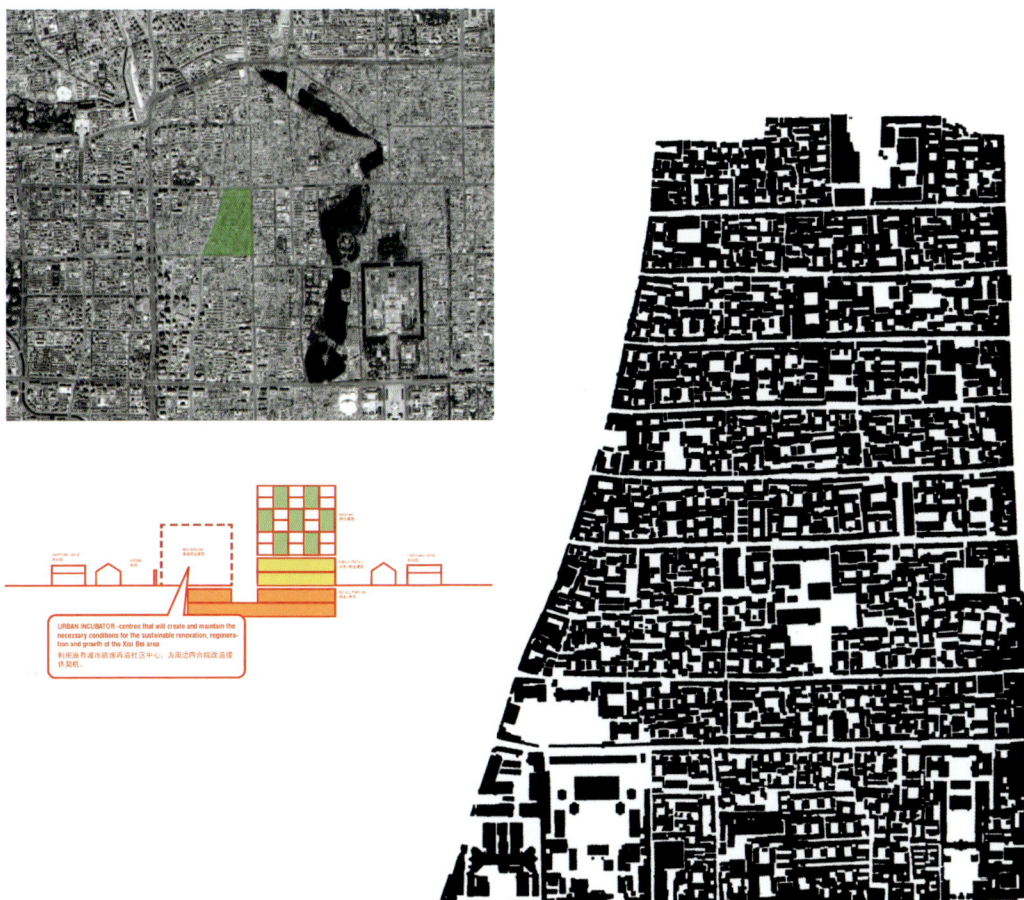

베이징의 전통 지역을 살펴보면 그 특징이 오로지 물리적인 형태에서 비롯되는 것이 아니라 그 지역에 살고 있는 주민들의 생활문화 영향이 크다는 것을 알 수 있다. 그 두 요소는 후통의 몸과 마음 같은 것이며 후통이 특별한 이유이기도 하다. 후통 지역을 재개발하기에 앞서 우리는 기존 주민과 그들의 생활 패턴, 인간관계라는 복잡한 네트워크를 어떻게 존속시킬지에 대해 고민해야 했다. 그것을 성공적으로 이뤄내는 동시에 생활환경을 개선하고 새로운 경제·사회적 원동력을 소개하기 위해 우리는 혼잡한 후통 지역 내에서 성장을 위한 공간을 찾아야 했다.

이 공간은 베이징 구심 여러 곳에 존재하는 '종양'에서 찾을 수 있었다. 이들은 1949년 이후 전통 지역을 침범한 큰 규모의 건물들로 주로 산업 또는 행정적인 기능을 수행하고 있었다. 후통의 리듬과 흐름을 방해하고 있는 이들은 현재 베이징이 안고 있는 더 큰 문제의 한 증상이지만 한편으로는 고밀도 다층 구조로 후통 지역의 재개발에 필요한 기회를 제공한다. 건축적 변화와 적절한 프로그램 소개로 우리는 이 '종양'을 '도시의 인큐베이터'로 탈바꿈시킬 수 있으며, 시쓰 베이 지역이 필요로 하는 지속 가능한 재생과 성장을 위한 장소로 기능할 수 있다. 우리는 이를 통해 변화와 재생이라는 후통의 전통을 현대 중국 도시 콘텍스트로 끌어올 수 있다.

The character of Beijing's traditional areas derives not only from their physical form, but perhaps more significantly from their inhabitants' culture of living - these two elements are the body and soul of the Hutongs, and are the essence of what makes them unique. In seeking to regenerate Hutong districts, we must therefore aim to retain existing populations, their living patterns and complex networks of relationships. If we are to achieve this, while simultaneously improving living conditions and introducing new economic and social impetus, then we need to find space for growth in the overcrowded Hutongs.

This space can be found in the numerous "tumors" to be found in Beijing's Old City. These are large-scale buildings introduced into traditional areas after 1949, and are usually industrial or administrative in function. Disruptive to the rhythm and flow of the Hutongs, these buildings represent a symptom of a wider urban problem, but they also represent an opportunity: They are existing, high density, multi-storey structures that can provide the space we need to regenerate the Hutongs. Through architectural transformations and the introduction of suitable programmatic elements, we can convert these tumors into "urban incubators": centres that will create and maintain over time the necessary conditions for the sustainable regeneration, renovation, and growth of the Xisi Bei area. In this way we can enable the Hutong tradition of transition and renewal to continue into the contemporary Chinese urban context.

WHILE SIMULTANEOUSLY IMPROVING LIVING CONDITIONS AND INTRODUCING NEW ECONOMIC AND SOCIAL IMPETUS, WE NEED TO FIND SPACE FOR GROWTH IN THE OVERCROWDED HUTONGS.

생활환경을 개선하고 새로운 경제·사회적 원동력을 소개하기 위해 우리는 혼잡한 후통 지역 내에서 성장을 위한 공간을 찾아야 했다.

A CHINESE PROPHECY ON FUTURISTIC ARCHITECTURE
미래주의 건축에 대한 중국의 예언

Zhou Rong 조우 롱(周榕)

세계 건축 역사는 혼란과 경쟁으로 가득한 소비 시대에 진입하고 있다. 오늘날 건축계는 경제 사회가 겪고 있는 에너지와 자원의 위기 등 유사한 문제점들을 안고 있다. 서양 문화의 이념들은 이제 시대에 뒤떨어진 것이 되었으며, 제한된 공급은 계속되는 서양식 개발에 에너지 부족이라는 결과를 초래했다. 유일한 자극제는 소비라는 촉매 위주로 집중된 일련의 실험뿐이다.

근대건축운동을 시작으로 건축가들은 마치 미래를 예측하는 양 예언자가 되었다. 미래 예측은 처음에는 현실 비판을 의미했지만 점차 '미래'의 대량생산으로 유토피아라는 상상을 만들었다. 그러다 유토피아의 꿈마저 사라지자 건축의 미래는 하나의 패션쇼가 되었고, 대중의 주의를 끄는 것은 유명 브랜드 간의 경쟁, 즉 메시지를 전달하기 위해 매체의 권력과 싸우는 것이 되었다. 원래의 예언자는 사라지고 가짜 예언자들이 훌륭한 '연기자'로 둔갑한 것이다.

이러한 국제무대에서 중국 건축가들은 그저 청중이었다. 30년 동안 지속된 고립과 따라잡아야 할 세월만 20년인 중국인들만이 새로운 메시지를 수용할 수 있었다. 그렇다면 그들은 어떻게 세계 건축의 미래를 내다볼 수 있을까?

중국의 개방과 빠른 경제성장으로 중국 건축가들은 마침내 세계 건축의 미래를 점쳐볼 기회를 얻었다. 중국을 대표하는 주 페이와 우 통은 프랭크 로이드 라이트, 프랭크 게리, 자하 하디드에 이어 구겐하임 미술관 설계를 의뢰 받은 네 번째 건축가가 되었다.

다른 중국 현대건축가들과 마찬가지로 주 페이 역시 서양 교육과 실무 경험에서 출발했다. 거의 10년 동안 외국에서 공부하며 실력을 쌓은 주 페이의 건축은 서양 건축문화의 언어를 기반으로 하고 있다. 중국으로 돌아와 작업한 초기 작품은 중국이나 지역 영향을 전혀 받지 않았으며 근대적이고 이상적이었다.

유명한 중국 디자이너 중 드문 여성 디자이너인 우 통은 그래픽디자인과 미술을 배경으로 건축에 발을 들인 경우다. 그녀는 디지털 베이징 작업으로 처음 주 페이와 협력했으며, 파사드 디자인은 미술과 디자인 업계를 잘 이해하는, 정보화시대의 미학에 대한 그녀의 개념에서 나왔다. 여러 분야와의 협력은 스튜디오 페이주의 창의력을 한층 향상시켰다.

디지털 베이징 이후 스튜디오 페이주는 세 가지 측면에서 생각의 진화를 통해 점차 변화된 모습을 보였다.

첫째, 디지털 베이징 프로젝트 이후 주 페이와 우 통의 디자인 비전은 순전히 건축적인 전통에서 보다 포괄적인 예술 세계로 넓혀졌다. 다른 예술 분야에서 영감을 받아 건축 디자인에 접목시키며 순수한 형태만을 고집하는 근대건축가적인 성향에서 복합적인 환경을 추구하는 신근대주의적 건축 성향으로 큰 변화를 보였다. 한편 현대예술 이념에 대한 이해는 주 페이와 우 통이 세계적인 명성을 얻은 건축가들과 경쟁하고 인정받을 기반을 마련해주었다.

둘째, 주 페이와 우 통은 그들의 디자인적 사고를 돕기 위해 현대 중국 도시를 포함한 도시적 환경으로 주의를 돌렸다. 그들은 중국과 아시아 도시의 매력과 활기는 그 표면적인 '더러움, 혼란 그리고 불화'에서 온다는 것을 알았다. 서구적인 개념의 도시 문제는 중국 건축가들에게 창의력을 선보일 훌륭한 기회가 되었으며, 닝보 북시티나 베이징 퍼블리싱 하우스 또는 베이징 시쓰 베이 재생과 같은 도시계획 프로젝트를 통해 도시의 활성화에 대한 주 페이와 우 통의 관심을 읽을 수 있다.

마지막으로 가장 중요한 점은 주 페이와 우 통이 중국의 문화적 정체성과 역사에 대한 인식을 깨우쳤다는 점이다. 서구적인 근대 건축문화로 제한되었던 그들의 디자인은 이제 중국 전통을 포용하고 있다. 과거와 단절된 채 미래를 바라보는 근대주의 이념과는 달리 중국인들은 과거, 현재, 미래의 관계를 중시하고 과거와 미래를 상반된 개념이 아닌 역동적이며 분리할 수 없는 하나로 이해한다. 중국인들에게 역사는 최후의 순간을 향해 선형으로 진행하는 서양의 믿음과는 반대로 영원토록 순환하는 것이다. 스튜디오 페이주의 최근 작품은 전통과 미래에 대한 그들의 새로운 생각과 정의를 표현하고 있다. 차이 귀치앙의 전통 가옥 개축과 강가의 자갈에서 영감을 얻은 위에 민췐 미술관에 표현된 시공간적 침투가 그 예다. 여기서 '역사'는 더 깊은 해석을 통해 비현실적인 상상적 미래가 아닌 과거의 성찰에서 비롯된 미래의 언어, 질감, 무게감 및 온도를 정의하고 있다.

서양 건축계에서 중국인이 서양 건축 언어만으로 건축의 미래를 예측한다는 것은 쉽지 않다. 중국의 현대건축은 과거로 돌아갈 수도 없고 서양의 미래주의가 될 수도 없다는 사실을 깨달은 주 페이와 우 통은 지혜롭다. 이와 같이 어두운 시기에 미래에 대한 중국의 예언을 다시 쓰는 유일한 방법은 중국 문화와 문명의 뿌리로 돌아가는 것이며, 그것만이 중국의 현대건축가들이 국제적인 무대를 배경으로 같은 수준에서 경쟁할 수 있는 유일한 길이다.

The history of global architecture is entering a consumption era, full by chaos and competitions. The realm of architecture is facing problems similar in nature to those that are ailing the world economies: a combined crisis in the energy and resources sectors. Ideologies that have fueled the Western cultural realm have gone old and limited in supply, resulting in a lack of energy, while maintaining the metabolism of Western development. The only stimulator is a series of experiments based on a high concentration of consumption catalysts.

Since the modern architectural movement, architects have eventually become prophets, pretentiously engaging in predicting the future. At first, the business of predicting the future meant making a criticism on reality. Later on, the mass production of the "future" was meant to consolidate an imagined utopia. When the utopian dream had vanished, predictions on the future of architecture became no less than a fashion show. What now catches people's attention is merely the competition between famous brands; the fighting with the power of the media so as to convey messages. Fake prophets have become better "actors," while the original prophets have disappeared.

In this international environment, Chinese architects were originally mere spectators. After thirty years of isolation from the outside world, with a resulting twenty years in order to play catch-up, the Chinese could only accept the new messages. How could they make predictions on the future development of world architecture?

Now, finally, Chinese architects have gained an opportunity to predict the future of world architecture, thanks to a rapid economic growth and to the opening of the country. Representing Chinese architecture, Zhu Pei and Wu Tong were the 4th invited to design a Guggenheim Museum, the only architects other than Frank Lloyd Wright, Frank Gehry, and Zaha Hadid to have this opportunity.

Like any other Chinese contemporary architect, Zhu Pei's starting point originates in his background of Western education and practice. Having studied and worked for almost ten years abroad, Zhu Pei has become well-versed in a language based in the Western architectural culture. His early works, after having moved back to China, have no reference to any local or Chinese influence. They are thoroughly modern and utopian.

Wu Tong, belonging to the female minority among famous Chinese designers, began to explore architecture from a background of graphic design and arts. Her first collaboration with Zhu Pei was 'Digital Beijing'. The facade design originates from her success in arts and the design industry, and from her ideas on aesthetics in the information age. The cross-disciplinary collaboration has made Studio Pei-Zhu more vibrant on a creative level.

Studio Pei-Zhu's works have shown an apparent change after the Digital Beijing project, due to three lines of development in their approach to design:

First of all, Zhu Pei and Wu Tong's design vision has moved from a pure architectural tradition to a broader, artistic one, since the Digital Beijing project. They took inspiration from forms in other arts and have integrated them in architectural design. Eventually they changed drastically, from the modern architects who insisted on and sought pure forms, to neo-modern architects who can create and co-exist with complex environments. In addition, their understanding of the world contemporary art ideologies guaranteed that Zhu Pei and Wu Tong are able to compete with the world's renowned architects and receive recognition.

Secondly, Zhu Pei and Wu Tong began to absorb ingredients from urban environments, especially those of contemporary Chinese cities, in order to support their design principles. They have clearly understood that the most attractive and vital parts of Chinese and Asian cities come from their superficial "dirtiness, chaos, and difference." These urban problems, as seen from a Western perspective, have actually become great creative opportunities for Chinese architects. From Ningbo Bookcity, to Beijing Publishing House, to urban design projects such as Beijing Xisi Bei Urban Regeneration, it is obvious that Zhu Pei and Wu Tong have an ambition in revitalizing and changing the city.

The last and the most important transformation has been Zhu Pei and Wu Tong's awakening to a Chinese cultural identity and historical consciousness. Their design vision has gradually moved from the narrow and modern Western architecture culture to a broader and deeper Chinese tradition. Different from Modernist ideologies, in which a futuristic style has a clear cut with the tradition, Chinese people are more interested in connecting to the past, present, and future, in a dynamic and inseparable whole, as opposed to the conceptual opposition between past and future. The Chinese perspectives on history (the cyclic, the endless, the eternal, and the co-existing) are in opposition to Western culture thought (the linear, the final, and the ultimate). Studio Pei-Zhu's recent works have shown their new philosophy and definitions of tradition and the future. This includes the penetration between time and space in Cai Guoqiang Courtyard House's renovation, and the Yue Minjun Art museum, whose inspiration comes from a river stone. This so-called "history" has achieved greater depth and interpretation. The defined semantics, texture, weight, and temperature of the future do not come from unrealistic imagination, but originate in a deep reflection of history.

Within the Western architecture realm, it is difficult for Chinese architects to fight for a voice needed to predict the future of architecture, using a pure Western architecture language. Zhu Pei and Wu Tong's wisdom comes from the fact that they have realized that Chinese contemporary architecture cannot return to the past, and it cannot go into the Western futuristic style. In this blind moment, the only way to rewrite the book of Chinese prophecy on the future is to reconnect back to the roots of Chinese culture and civilization. This is the only way for Chinese contemporary architects to catch up and compete on an international level.

조우 롱은 베이징 칭화대 건축학과 부교수로 건축 비평가이다.
Zhou Rong is an associate professor in School of Architecture, Tsinghua University, Beijing, and an architecture critic.

YOUNGFU
TURISMEX
TREMEUN
REALISTIC
UNIVERSA
IMPLICA

STUDIO
MAD

매드(MAD)는 현대건축, 도시계획, 문화 분석을 중심으로 작업하는 젊은 건축가 집단이다. 베이징을 거점으로 오늘날의 중국과 세계의 사회학, 테크놀로지, 정치 등 다양한 영역을 탐구하며 건축의 경계를 넘어 미래주의의 개념을 연구 중이다.

MAD is a young but innovative architectural design office practicing contemporary architecture, urbanism and cultural analysis. Based in Beijing, we examine and develop our concept of futurism beyond the boundaries of architecture, by exploring into sociology, technology and politics in today's China. MAD's ongoing projects include: the Absolute Tower in Toronto, Canada, an international competition won by MAD in 2006; the Tianjin Sinosteel International Plaza, a 358M high-rise building in Tianjin, China; the Mongolian Museum in Inner Mongolia, China; and large-scale public complexes and residential housing in Denmark, Hong Kong, Dubai, Singapore, Malaysia, Japan and Costa Rica. MAD's work has been published worldwide. In 2006, MAD was awarded the Architectural League Young Architects Forum Award. The office has also presented a series of exhibitions, including the "MAD in China" exhibition at the Venice Architecture Biennial, and the "MAD Under Construction" exhibition at the Beijing Tokyo Art Projects Gallery in Beijing. In 2007, "MAD in China," a floating city of MAD's work, was shown at the Danish Architecture Centre in Copenhagen, Denmark.

베이징 출신인 마 옌송은 2002년 예일대학교 건축대학에서 건축 석사 학위를 받았으며 런던의 자하 하디드 아키텍츠와 뉴욕의 아이젠만 아키텍츠에서 프로젝트 디자이너로 경험을 쌓은 뒤 2004년 매드를 설립했다. 베이징중앙미술대학교에서 건축을 가르쳤으며, 2006년 뉴욕 아키텍처 리그의 '젊은 건축가상'과 2002년 사무엘 J. 포젤슨 우수 디자인상을 수상했고, 2001년 AIA로부터 건축 연구를 위한 장학금을 받았다. 2006년 토론토에서 열린 앱솔루트 타워 공모전, 2005년 중국 광저우에서 열린 솔라 플라자 공모전, 2004년 상해 국립 소프트웨어 아웃소싱 베이스 등 여러 국제 설계경기에 당선되었다. 그의 작품 '세계무역센터 재건 – 떠 있는 섬'과 '수조'는 베이징 건축 비엔날레와 2004년 중국 국립미술관에 전시되기도 했다. 또한 그의 설치작품인 '잉크 아이스'는 2005 중국 서예미술전에 소개되었으며, 2006년 매드는 베니스 건축비엔날레에 〈매드 인 차이나〉라는 제목의 전시로 소개되었다. 베이징 도쿄 갤러리에서 〈매드 공사 중〉전을 열었으며, 2007년 11월 3일부터 2008년 1월 6일까지 덴마크 코펜하겐의 덴마크 건축센터에서 〈매드 인 차이나〉전이 열렸다.

일본 나고야 출신인 요스케 하야노는 2000년 도쿄 와세다대학에서 재료공학 학사 학위를 취득한 뒤 이듬해 와세다 예술건축대학에서 건축 준학사 학위를 취득했다. 2003년 런던 AA스쿨에서 전문가 석사 과정인 DRL을 마쳤으며, 그의 졸업 프로젝트인 '소호텔/시냅스'는 2002년 프랑스 오를레앙에서 열린 국제 건축 컨퍼런스 아키랩과 오스트리아 그라츠에서 열린 〈잠재적인 유토피아〉전에서 소개되었다. 하야노는 2006 뉴욕 아키텍처 리그의 젊은 건축가상을 수상했다. 매드에 합류하기 전 런던에 있는 자하 하디드 아키텍츠에서 프로젝트 디자이너로 경력을 쌓았으며, 현재 와세다 예술건축대학에서 강의 중이다.

당 췬은 상해 출신으로 아이오와 주립대학에서 건축 석사 학위를 받았다. 매드에 합류하기 전 퍼킨스 이스트만 등 미국의 여러 주요 건축설계사무소에서 다양한 스케일의 프로젝트를 통해 경력을 쌓았다. 아이오와 주립대학의 조교수, 프랫 인스티튜트의 객원교수, 이탈리아 로마에서 진행되는 아이오와 주립대학의 해외 학습 프로그램 조교수 등으로 교육 활동에 참여했으며, 여러 전문지와 국내 건축전 및 컨퍼런스 등을 통해 글과 작품을 소개했다. 당 췬은 2000년 AIA의 우수상을 수상했으며, 2001년 폴 S. 건축문화상과 디자인 미디어상, 아이오와 주립대학의 우수상 등을 수상했다.

MA YANSONG

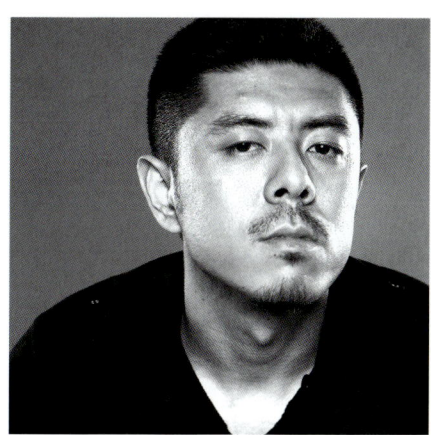

FOUNDING PRINCIPAL 마 옌송(설립자 및 대표)

Ma Yansong, originally from Beijing, received his Master of Architecture from the Yale University School of Architecture in 2002. Prior to founding MAD in 2004, Mr. Ma worked as a project designer with Zaha Hadid Architects in London, and with Eisenman Architects in New York. He also taught architecture at the Central Academy of Fine Arts in Beijing. Ma was the winner of the 2006 Architecture League of New York Young Architect Award. He also received the 2002 Samuel J. Fogelson Memorial Award of Design Excellence, as well as the American Institute of Architects Scholarship for Advanced Architecture Research in 2001. His works have won numerous international design competitions, including: the 2006 Absolute Tower Competition in Toronto; the 2005 Solar Plaza Competition in Guangzhou, China, and the 2004 Shanghai National Software Outsourcing Base. Two of his works, WTC Rebuilt - Floating Island, and Fish Tank were exhibited at the Beijing Architectural Biennale and featured at the National Art Museum of China in 2004. His art installation 'Ink Ice' was featured in the Chinese Calligraphy Art exhibition in 2005. In 2006, MAD was shown at the 'MAD in China' exhibition in Venice during the Architecture Biennial, and the 'MAD Under Construction' exhibition at the Tokyo Gallery in Beijing. From Nov. 3, 2007 until Jan. 6, 2008, "MAD in China," a floating city of MAD's work, has been shown at the Danish Architecture Centre in Copenhagen, Denmark.

YOSUKE HAYANO

PRINCIPAL 요스케 하야노(대표)

Yosuke Hayano, originally from Nagoya, Japan, received his Bachelor of Materials Engineering from Waseda University in Tokyo in 2000, his Associate degree in Architecture from Waseda Art and Architecture School in 2001, and his Post-Professional Master of Architectural Design from the Design Research Laboratory of the Architectural Association of London in 2003. His thesis project SoHotel/Synapse was exhibited at Architlab, the International Architectural Conference in Orleans, France in 2002; and at 'Latent Utopias' in Graz, Austria in 2002. Mr. Hayano was the winner of the 2006 Architecture League of New York Young Architect Award. Prior to MAD, Yosuke Hayano was a project designer for Zaha Hadid Architects in London. He is currently a visiting lecturer at the Waseda Art and Architecture School.

DANG QUN

PRINCIPAL 당 췬(대표)

Dang Qun, originally from Shanghai, received her Master's Degree in Architecture from Iowa State University. Prior to MAD, Dang Qun worked for several major architecture firms in the United States, including Perkins Eastman, on projects of different scales. She has had several teaching positions, including: an assistant professorship at Iowa State University, a visiting professorship at Pratt Institute, and an assistant professorship at Iowa State University's Foreign Studies Program in Rome, Italy. Her writings and projects have been published in several professional journals; and her projects have been exhibited in national architectural exhibitions and conferences. She received the 2000 American Institute of Architect's Certificate of Merit. She also received the Paul S. Studies in Architecture & Culture Award, the Design Media Award and the Academic Excellence from ISU in 2001.

Hongluo Club

홍 루 오 클 럽

Beijing, China

Director in charge: Ma Yansong, Yosuke Hayano **Competition team:** Florian Pucher, Shen Jun, Christian Taubert, Marco Zuttioni, Yu Kui **Associate architects and engineers:** IDEA Design Studio **Building area:** 189.7m² **Completion:** 2006.5 **Client:** Beijing Earth Real Estate Develops Company **Photo by:** Shu He, Sun Xiangyu

베이징의 확장은 지난 몇 년간 변두리 지역의 개발에 박차를 가했다. 이는 중국의 첫 번째 공간 유형의 발달을 도왔으며 점차 교외 발달로 이어졌다. 나무다리를 통해 진입하는 클럽하우스는 호수에 떠 있는 수영장과 수중 플랫폼으로 이루어져 있다. 건물 형태는 사용자의 동선을 반영하고 있는데 클럽하우스 중앙에서 만나는 2개의 주요 동선은 상승하는 지붕을 따라 올라간다. 계속해서 변화하는 수면 또한 상승하는 지붕과 만나 액체에서 고체로의 변화를 표현하고 있다. 공간의 구조와 기능은 자연스럽게 통합되어 있다. 주 진입은 방문객을 수면 1,300mm 아래로 유도해 호수 한가운데를 걷고 있는 느낌을 준다. 클럽하우스에 다가갈수록 점차 지면과 같은 레벨로 올라오면서 집합의 장소라는 건물의 주요 기능을 드러낸다. 지붕 형태는 1층 프로그램의 선형적이고 기능적인 구조를 표현하며, 야외 수영장은 호수 중앙에 두어 자연 수면과 인공 수면이 같은 레벨을 갖게 했다. 건축을 통해 도시민의 자연에 대한 이해를 탐구하고 있는 홍루오 클럽하우스는 주변 환경에 반응해 계속해서 변화하는 공간으로 인간과 자연의 하나 됨을 시도하면서 산과 물을 통해 콘크리트 숲에 살고 있는 도시민에게 희망과 영감을 전달한다.

Photographs by Shu He

The expansion of Beijing City has intensively accelerated the development of its periphery area for the last few years, which has thereby helped develop the first space typology in China. In the meantime, the suburbia has emerged on a gradual scale. A wood bridge was introduced as an access to the Club House. The house has two branches: One is a swimming pool floating on the lake, the other is an underwater platform. The architectural form is shaped by people's circulation. Two major roads converge at the centre of the house, and reach all the way up, along an ascending roof. The ever-changing water surface joins the ascending roof, expressing the transition from liquid to solid. The space structure and the functions of the house are integrated naturally. The main access to the house will bring the visitor to 1,300mm under water, where people will feel like walking inside the lake. The access road gradually ascends to the ground level as it nears the house, which reveals the main function of the building - a gathering space. The roof shape is a projection of the linear, functional organization of the ground level's program. The outdoor swimming pool is built into the lake, which keeps the surfaces of the natural and the artificial water at the same level. The architecture explores the city dwellers' understanding of the nature. Hongluo Club House creates an ever-changing space that echoes with the surroundings, where people and the nature are united. The mountain and the water provide hope and inspiration for people who live in the concrete woods.

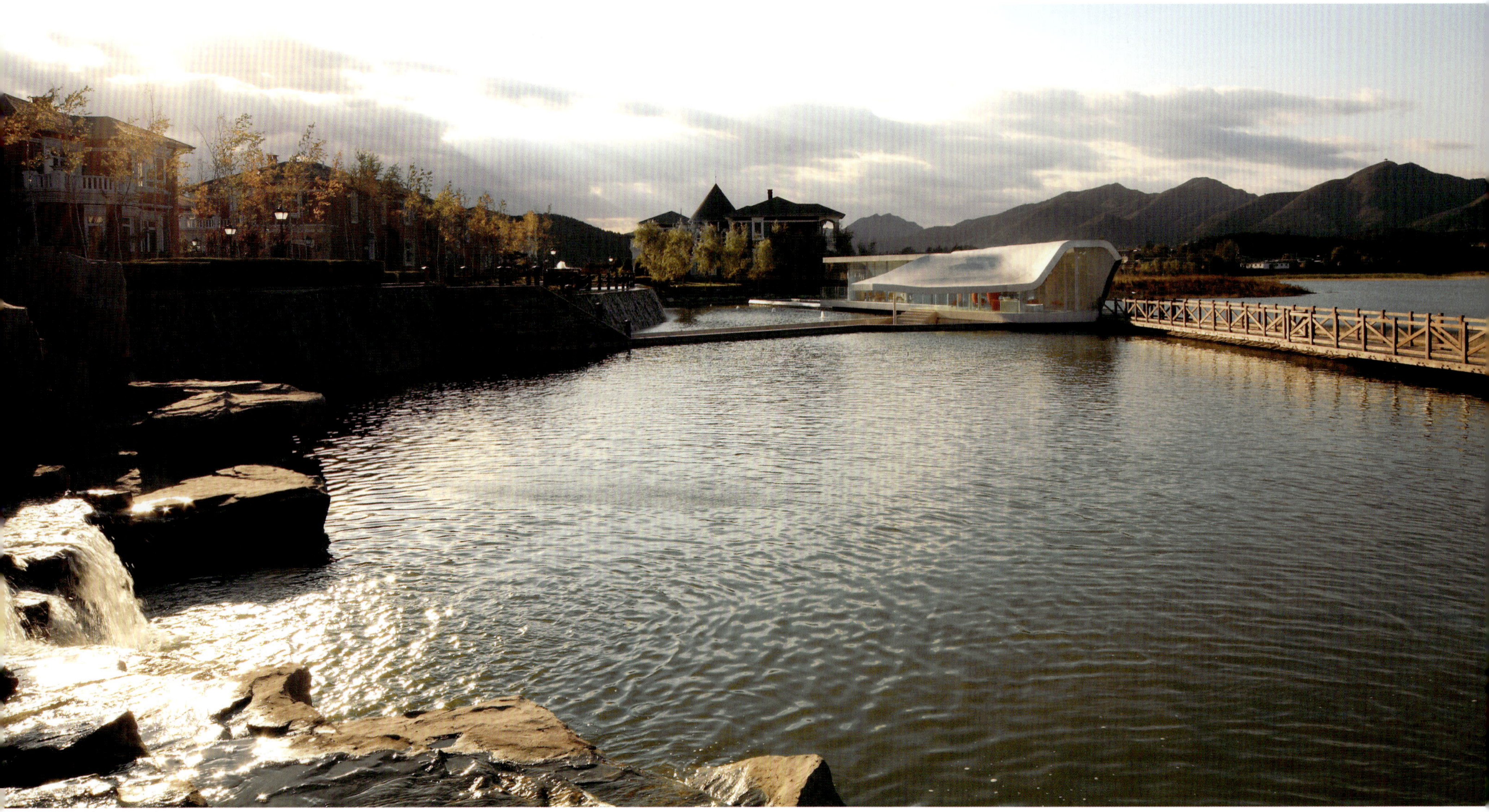

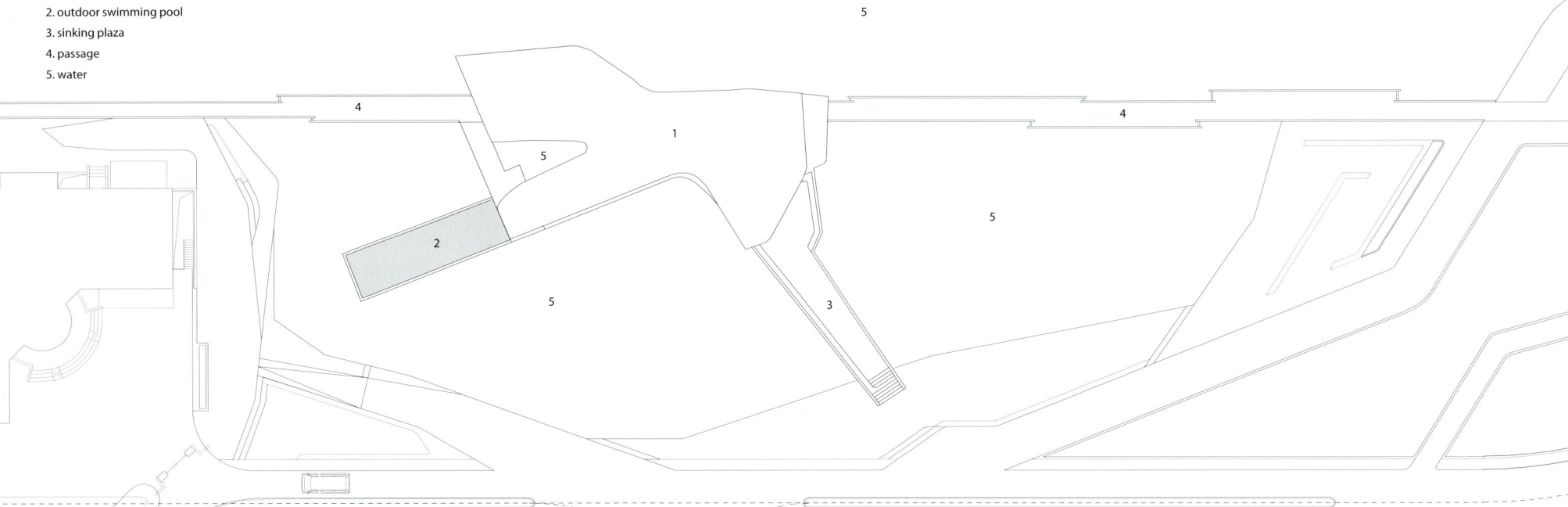

1. club
2. outdoor swimming pool
3. sinking plaza
4. passage
5. water
5
4
4
1
5
2
5
5
3
N

Master Plan

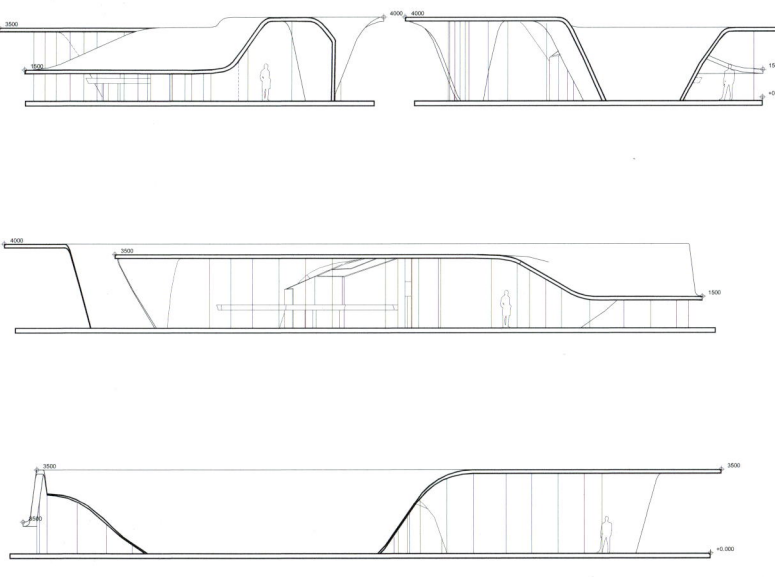
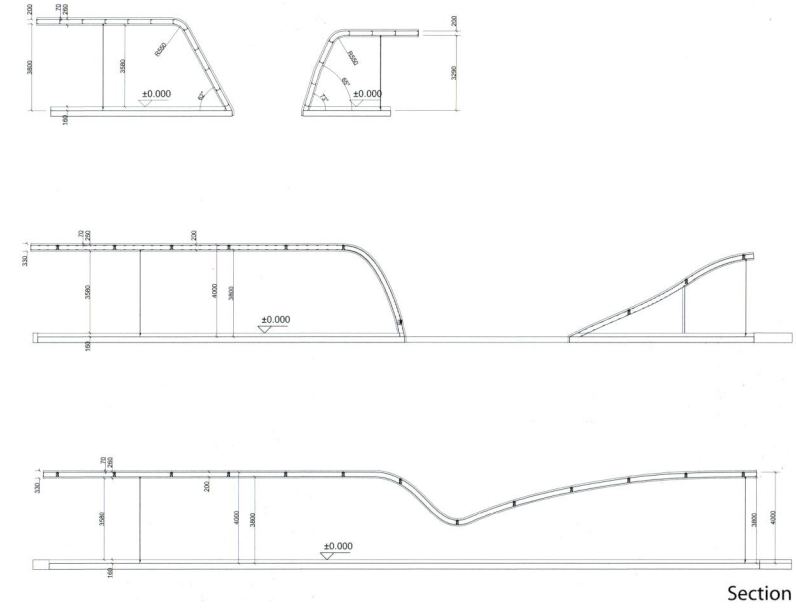

Section

A WOOD BRIDGE WAS INTRODUCED AS AN ACCESS TO THE CLUB HOUSE. THE HOUSE HAS TWO BRANCHES: ONE IS A SWIMMING POOL FLOATING ON THE LAKE, THE OTHER IS AN UNDERWATER PLATFORM.
나무다리를 통해 진입하는 클럽하우스는 호수에 떠 있는 수영장과 수중 플랫폼으로 이루어져 있다.

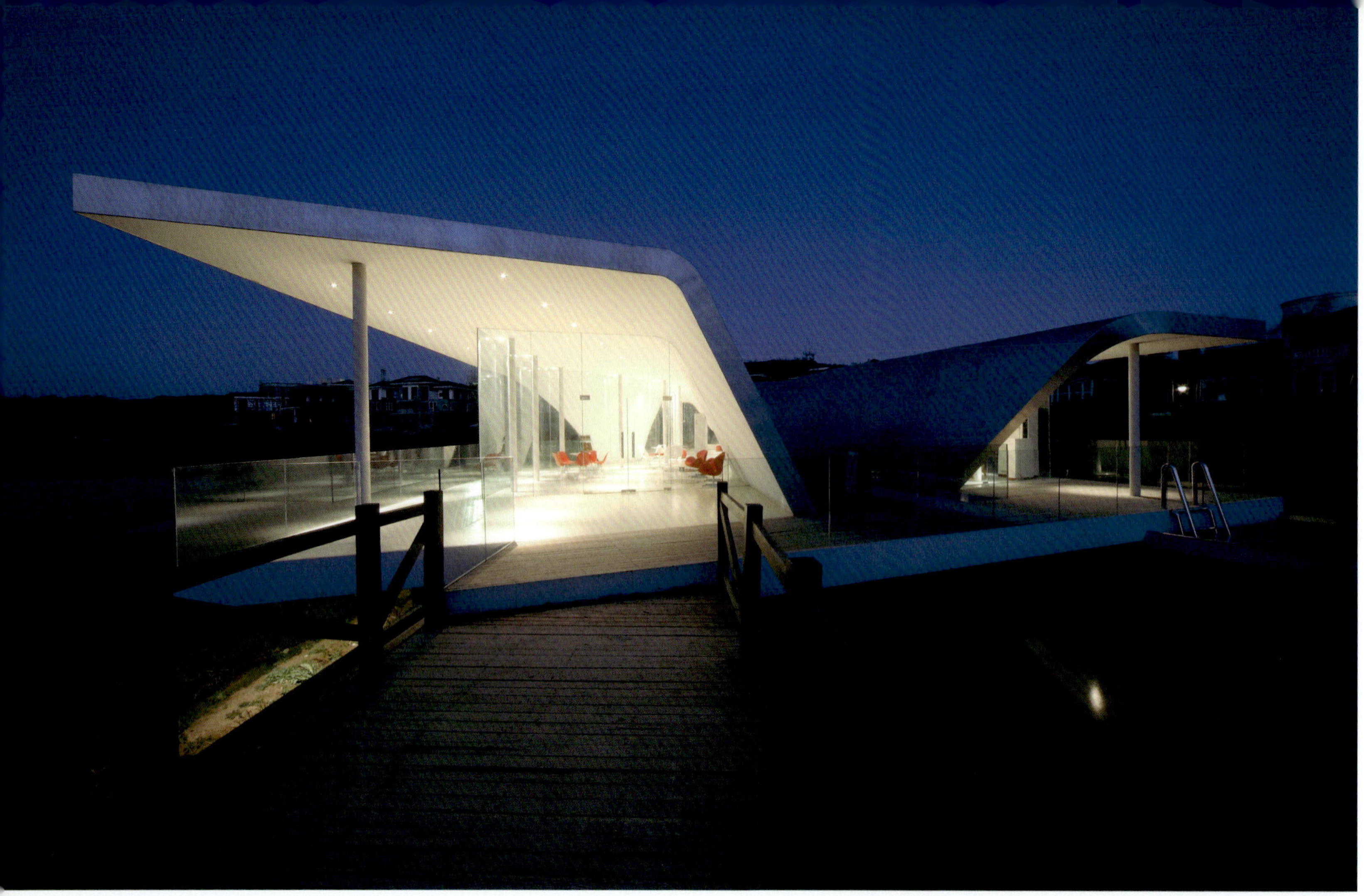

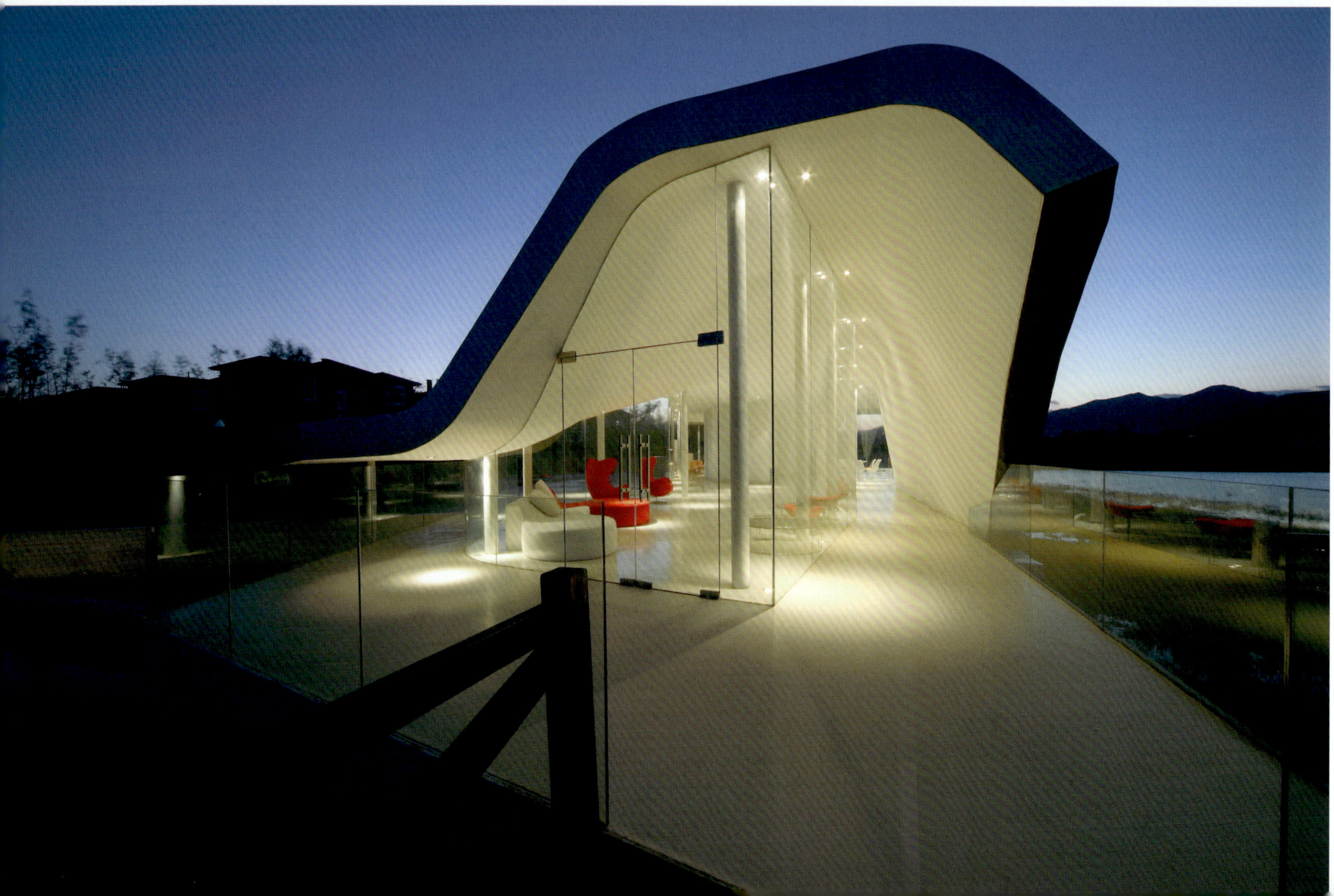

TWO MAJOR ROADS CONVERGE AT THE CENTRE OF THE HOUSE, AND REACH ALL THE WAY UP, ALONG AN ASCENDING ROOF.
클럽하우스 중앙에서 만나는 2개의 주요 동선은 상승하는 지붕을 따라 올라간다.

도 쿄 갤 러 리 B T A P 개 축

TOKYO GALLERY BTAP RENOVATION

Beijing, China

Director in charge: Ma Yansong, Yosuke Hayano **Design team:** Shen Jun, Emilio Doporto, Chiristine Yogiaman **Programme:** Art Gallery Renovation **Building area:** 364m² **Renovation area:** 136m² **Client:** Beijing Tokyo Art Projects (BTAP) Gallery

스틸, 유리, 콘크리트는 1950년대 독일의 건축운동 바우하우스의 영향으로 현대화의 상징이 되었다. 대량 생산된 재료는 건물의 양과 질을 결정하며 새로운 건물 유형을 탄생시켰고, 건축이 대지보다 공장을 더 의존하기 시작하면서 공장은 변화하는 도시의 기념물이 되었다. 탈근대시대 베이징에서 이와 같은 기념물들은 더 이상 사용되지 않는 빈 공장으로 남았고, 세계화라는 새로운 과제 아래 도시는 비어 있는 공간들이 새로운 의미의 공간으로 재탄생할 수 있는 잠재력을 발견했다. 21세기의 세계적인 대도시 베이징에 자리한 이 공장들은 예술 중심 지구로 변했다. 도쿄 갤러리는 20세기 말 798 예술지구에 문을 연 첫 갤러리로 베이징에 21세기 아시아 예술계를 위한 발판을 마련하려는 목표를 세웠다. 갤러리의 지난 몇 년의 성공에 이은 이 프로젝트는 갤러리의 전시 공간과 사무 공간을 통해 새로운 현대 공간을 표현하고자 하는 계획을 가지고 있다. 혁신은 새로운 테크놀로지에서만 나오는 것이 아니다. 혁신은 사용의 재발견에서도 찾을 수 있으며, 이 프로젝트에 주어진 주요 문제는 앞선 기술로 재료의 잠재력을 깨워 딱딱함과 부드러움 사이의 공간을 만드는 것이었다. 유리 박스로 된 사무 공간과 중층에 매달린 단단한 스틸 계단으로 이루어진 딱딱한 공간은 관람객의 주의를 끄는 첫 단계로 주요 전시 공간을 내려다볼 수 있는 중층으로 사람들을 이끈다. 계단의 스틸 조각들은 점차 부드러운 패턴으로 변하며 위층에서는 일련의 가구로 변신한다. 이와 같은 재료의 비선형적인 변화를 통해 재료와 공간에 대한 사람들의 선입견을 바꾼다.

Steel, glass, and concrete had become a symbol of Modernization in the 50s, influenced by the German architectural movement led by Bauhaus. Mass-produced materials provided a standard for quantity and quality of building, and created a building prototype. Building started to depend on factories rather than site; therefore, factories became monuments to city transformation. In the post-Modernization era, these monuments are left over as unused factories in Beijing. With a new agenda rooted in globalization, cities have found the potential to create new meanings for these spaces, as blank, flexible spaces. In the 21st Century, the cluster of factories will be transformed into a central art district in one of the world's biggest metropolis. Tokyo Gallery was one of the first galleries to open in the 798 Art District, just before the arrival of the 21st Century. Their aim was to anchor their art direction in Beijing, and to create a platform for the next century Asian art scene. After successful years in their gallery activities, this renovation project is planned to extend the gallery's possibility as exhibition space, as well as to alter the existing office space to express a new, modern space. Innovation comes from not only new technology, but also from re-evaluation of usage. The main challenge of the project is to find new potential for materials using advanced technology to create spaces between hard and soft. Hard contact between the Glass Box of office space and the solid steel staircase, which hangs from the mezzanine level, is the first focal point of the space which leads visitors to the mezzanine level, from where people have new viewing perspectives to the art pieces exhibited in the main space. The steel strips of the staircase are transformed to fluid patterns, and when it reaches the upper level, it becomes a series of successive furniture pieces. This nonlinear materials transformation changes people's presumptions on the relationship between materials and space.

Hard contact between the glass box of office space and the solid steel staircase, which hangs from the mezzanine level, is the first focal point of the space which leads visitors to the mezzanine level.

유리 박스로 된 사무 공간과 중층에 매달린 단단한 스틸 계단으로 이루어진 딱딱한 공간은 관람객의 주의를 끄는 첫 단계로 주요 전시 공간을 내려다볼 수 있는 중층으로 사람들을 이끈다.

MEZZANINE 1
CHAIR
RESTROOM
SLOPE 1%
MEZZANINE 2
STORAGE
15449

MAIN BEAM
LOOKOUT POINT 1
MEZZANINE 1
GLASS DIVISION
I-BEAM
DOOR CONECTION
OFFICE ACCESS
STEP 7
DOOR HANDLE
RESTROOM ACCESS
MAIN STAIR
DOOR LOCK
7084

MAIN BEAM
LOOKOUT POINT 1
MEZZANINE 1
STEEL PANEL
I-BEAM
MAIN BEAM
OFFICE
RESTROOM ACCESS
7084

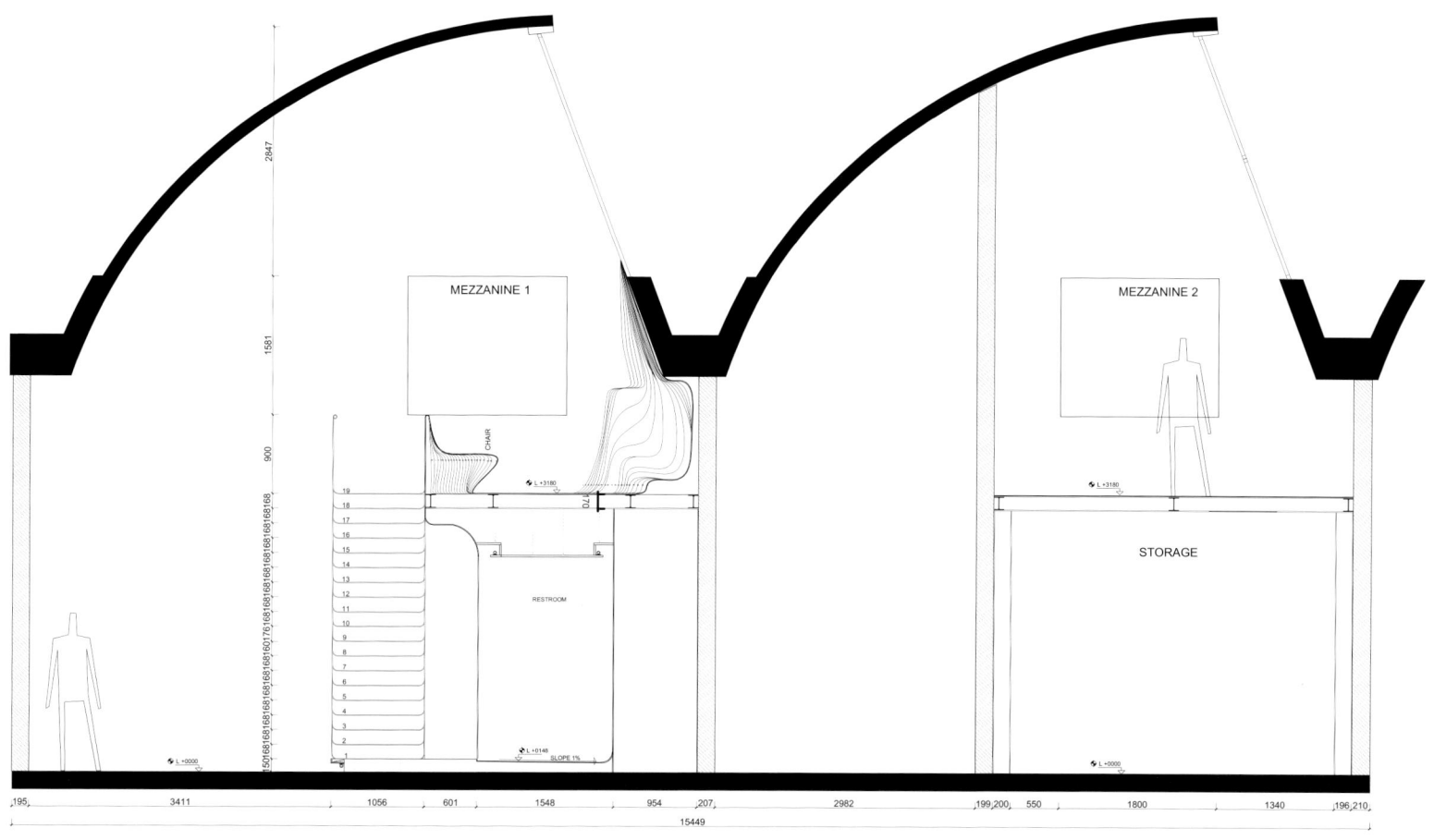

Section

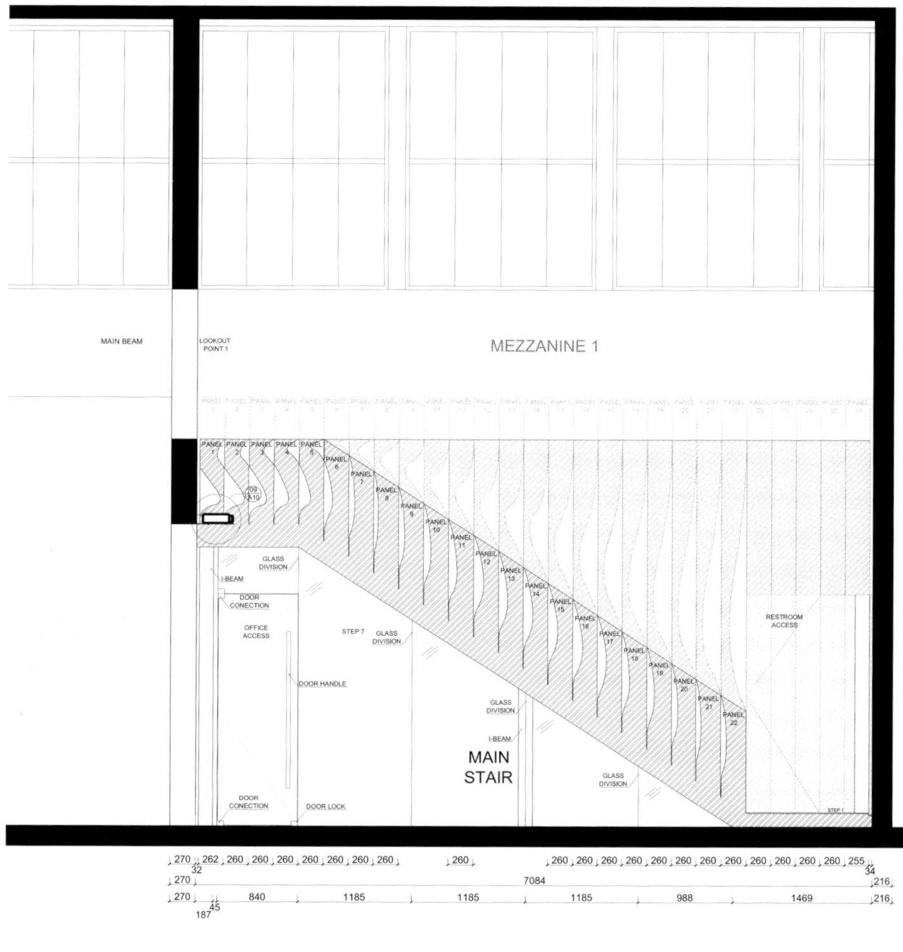

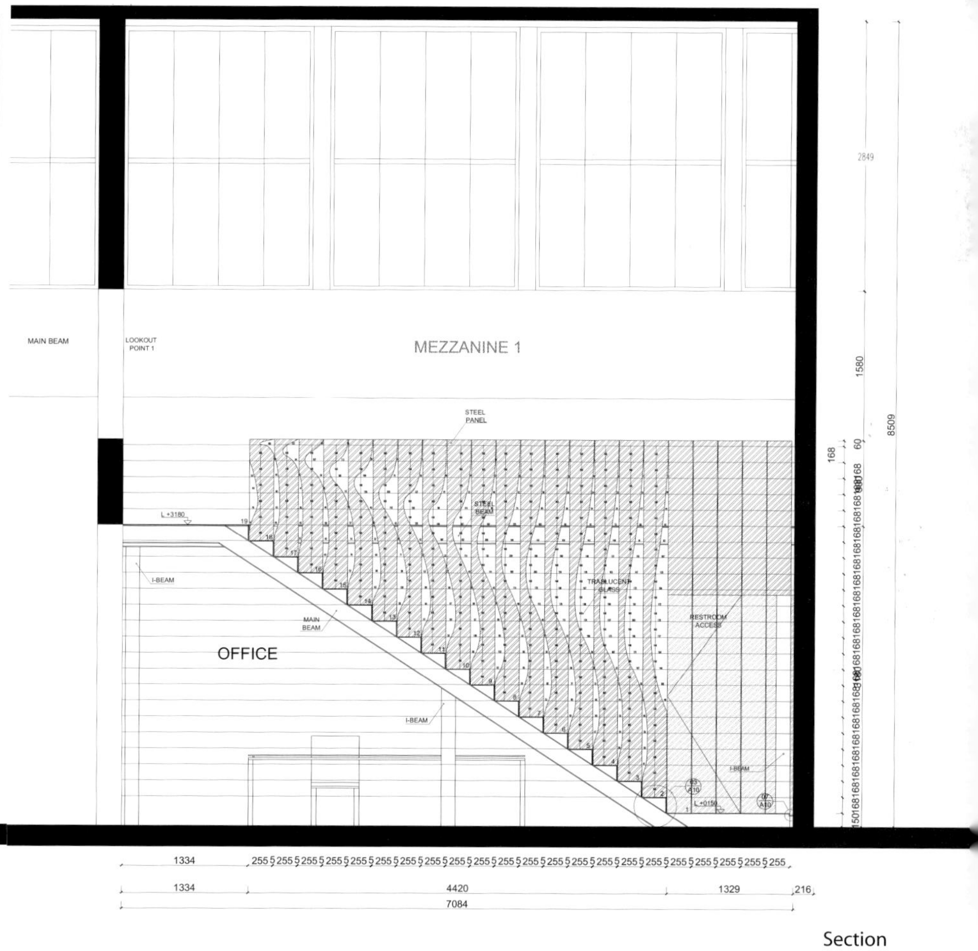

THE ABSOLUTE TOWERS
Mississauga, Canada

Director in charge: Ma Yansong, Yosuke Hayano, Dang Qun **Competition team:** Shen Jun, Robert Groessinger, Florian Pucher, Yi Wenzhen, Hao Yi, Yao Mengyao, Zhao Fan, Liu Yuan, Zhao Wei, Li Kunjuan, Yu Kui, Max Lonnqvist, Eric Spencer **Building area:** Phase 4 - 45,000m², Phase 5 - 40,000m²
Use: Residential Apartments **Building height:** Phase 4 - 56 stories (170m), Phase 5 - 50 stories (150m)
Completion: 2009 **Client:** Fernbrook Homes and Citizen Development Group

"집은 살기 위한 기계다"라는 유명한 근대주의 표어가 있다. 하지만 기계와 그로 구성된 사회가 큰 변화를 겪은 오늘날 우리는 건축을 어떻게 이해해야 할까? 건축이 산업시대로부터 멀어지고 있다면 건축이 전하는 메시지는 무엇인가? 북미 지역에서 빠르게 발전 중인 여러 교외 지역과 마찬가지로 미시소거 역시 자신의 특징을 잘 표현하는 새로운 정체성을 찾고 있다. 우리는 특별한 방식으로 성장을 희망하는 도시의 요구를 훌륭한 기회로 보고 대도시로 성장하려는 다른 작은 도시들의 사례를 그대로 따르기보다는 미시소거만의 특징을 살려야 한다고 생각했다. 우리의 설계안은 단순화라는 근대주의 원칙을 버리고 다양한 접근 방식을 통해 현대사회의 복잡성과 다양성을 보다 높은 수준으로 표현하는 한편, 여러 사회적 요구를 다양한 방식으로 수용하고 있다. 앱솔루트 타워는 디자인적인 저력을 과시하고 있을 뿐 아니라 주변 지역과 사회적 콘텍스트에 하나의 메시지를 전달하고 있으며, 후론타리오가와 범햄소프(Burmhamthope)가가 교차하며 타워에 자리한 미시소거시티센터로 들어가는 입구를 형성한다. 조형성이 강한 이 건물은 대담성, 관능성, 로맨스라는 보편적인 언어를 표현하고 있으며, 인간의 몸을 연상시키면서 비틀리는 리듬으로 미시소거의 새로운 랜드마크적 아이콘이 될 것이다. 건물 전체를 둘러싸고 있는 연속적인 발코니는 전통적으로 높이를 강조하기 위해 고층 빌딩에 쓰인 수직적인 선을 없앤다. 건물은 매 층 다른 각도로 회전하며 각자의 높이가 가진 풍경에 반응하고 있다. 우리는 도시민의 자연에 대한 꿈을 일깨워 태양과 바람을 느낄 수 있는 곳을 만들고자 했다. 지역 주민들은 앱솔루트 타워에 마릴린 먼로라는 별명을 붙였다.

Modernism has a famous motto: A house is a machine for living in. However, as we increasingly leave the machine age behind, we are left with a question: what message should architecture convey? What is a house for now? Like other fast developing suburbs in North America, Mississauga is seeking a new identity. This is an opportunity to respond to the needs of an expanding city, to create a residential landmark that moves beyond simple efficiency to provide an emotional connection for the people who live there. In place of the simple, functional logic of modernism, our design expresses the complex and multiple needs of contemporary society. This building is more than just a functional machine: it responds to its location at the junction of two main streets to become a landmark, a gateway that signifies entrance to a city. It is something beautiful, sculptural and human. Despite its landmark status, the emphasis is not solely on height. In our design, a continuous balcony surrounds the whole building, eliminating the vertical barriers that are traditionally used in high rise architecture. The entire building rotates by different degrees at different levels, corresponding with the surrounding scenery. Our aim is to provide 360 degree views, to awaken the city dweller's appreciation of nature, and to get them in touch with the sunlight and the wind. The Absolute Tower has been nicknamed Marilyn Monroe by the locals.

In our design, a continuous balcony surrounds the whole building, eliminating the vertical barriers that are traditionally used in high rise architecture.

건물 전체를 둘러싸고 있는 연속적인 발코니는 전통적으로 높이를 강조하기 위해 고층 빌딩에 쓰인 수직적인 선을 없앤다.

Section

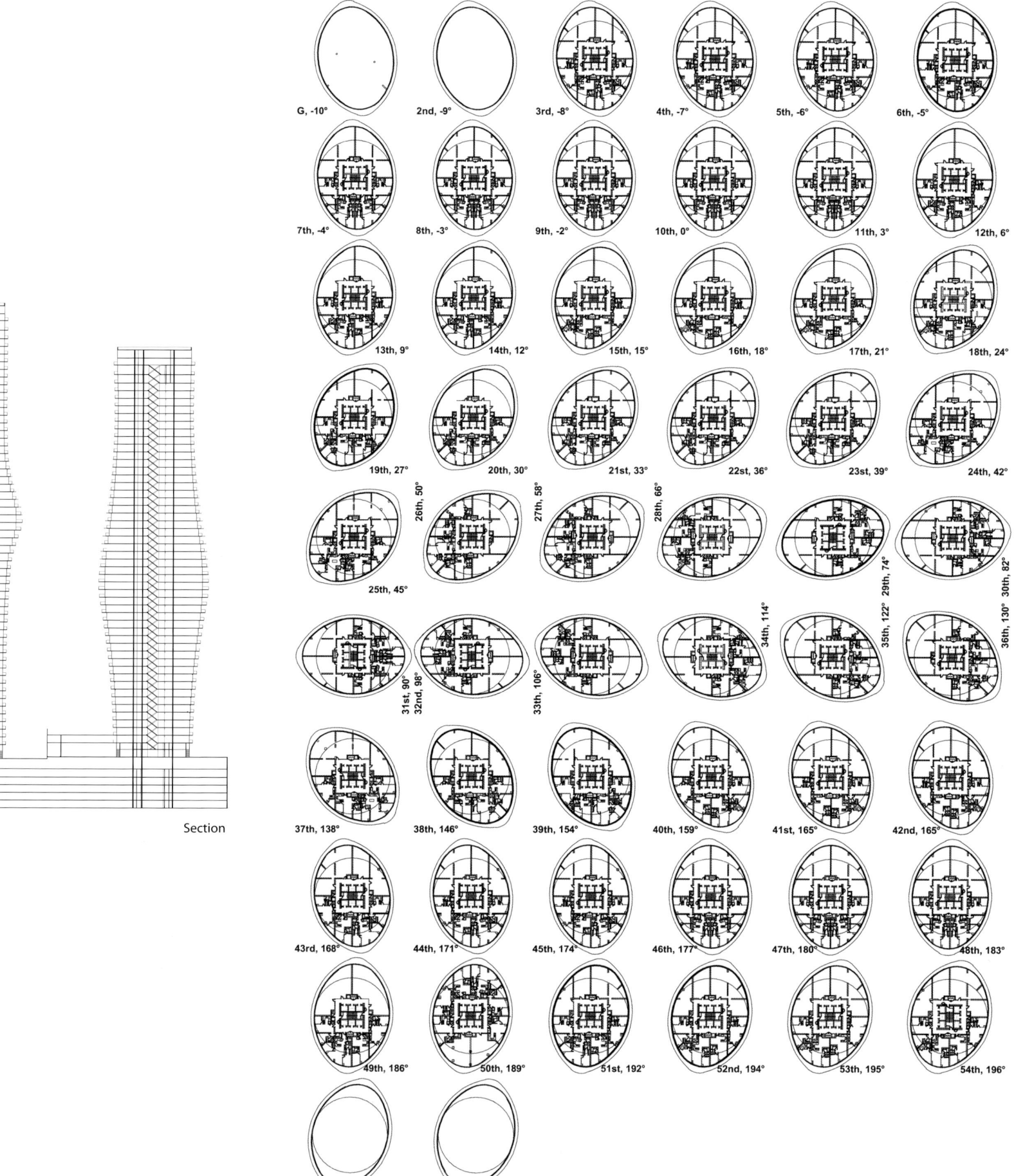

Diagram

Changsha Culture Park

창샤 문화 공원

Hunan, China

Director in charge: Ma Yansong, Yosuke Hayano, Dang Qun **Associate architects and engineers:** BIAD
Program: Museum, Opera House, Library **Building Area:** 60,000m² **Client:** Changsha Culture Government

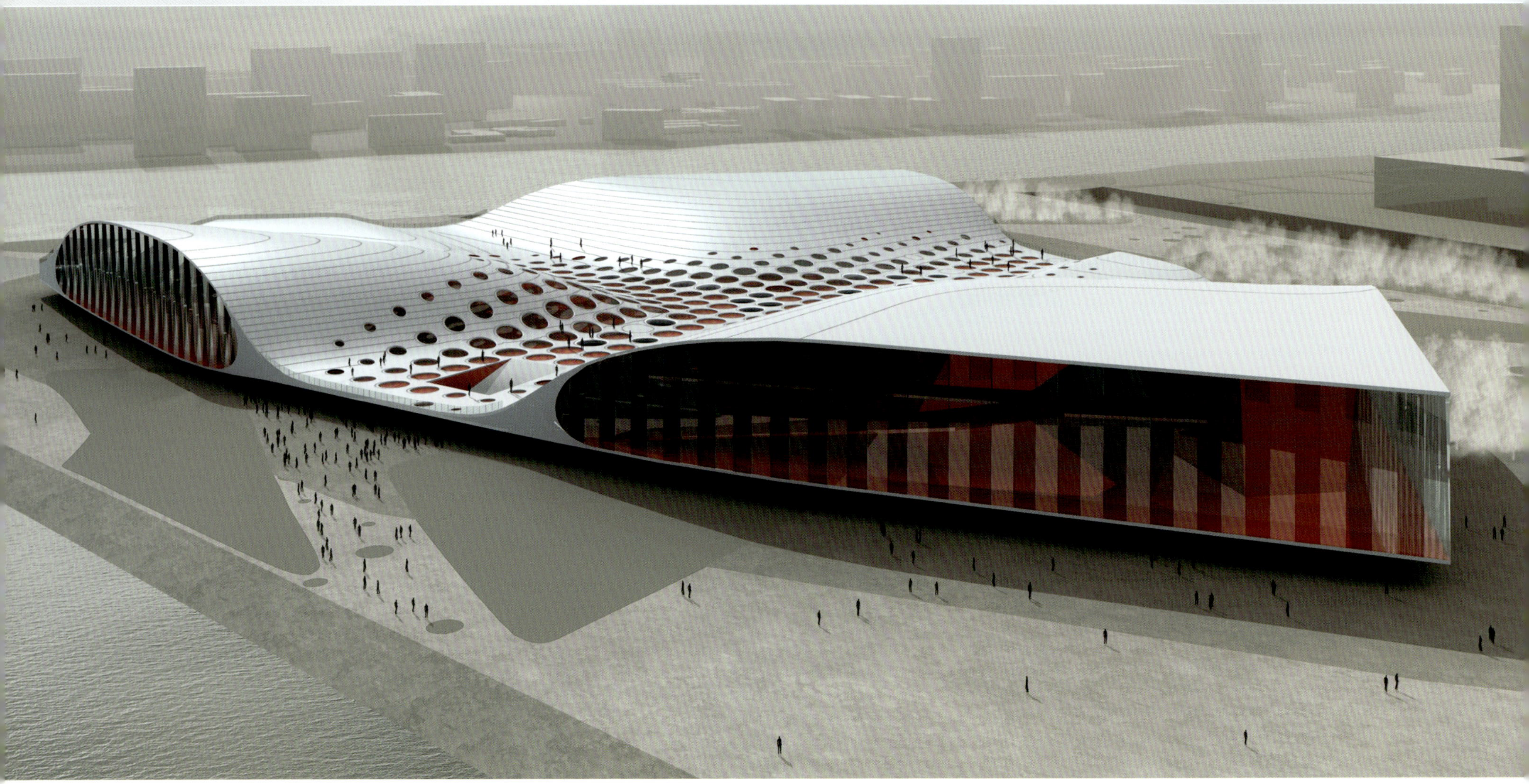

프로젝트의 가장 큰 주안점은 도시환경과 워터프런트를 유기적으로 연결하면서 창사시의 문화적 공간을 창출하는 것이다. 이 새로운 도시 고원은 내부 기능에 반응하고 통합하는 두 개의 표면을 가지고 있다. 외부 환경은 그 동안의 도시 발달과 함께 많은 변화에 시달렸지만 위대한 자연을 느낄 수 있는 축복받은 새 도시 공간이다.
새로운 문화공간이 만들어 내는 곡면 아래 음악당, 미술관, 도서관이 다 함께 자리하고 있다. 워터프런트의 경관은 이 삼차원의 공간에 또 다시 출현한다. 세 개의 프로그램은 문화공원의 이미지를 향상시키며 창사 시에 또 다른 매력을 부여한다.
이곳에서 사람들은 건물 안팎의 여러 층에 걸쳐 마련된 문화 시설들을 언제든지 즐길 수 있다. 전시는 물론 한낮의 여흥과 저녁의 축제에 이르기까지 이곳 문화공원에서는 다양한 대규모 활동을 펼칠 수 있다.

The biggest challenge of this architectural project is to create a new cultural plateau for Changsha City, which simultaneously serves as an organic link between urban context and waterfront. This new urban plateau has two surfaces that respond and articulate to internal function. The external environment turns out to be new urban space blessed with the richness of the great nature, which has been given a vigorous touch by the progression of the city. Changsha new culture platform creates a soft surface, under which the Music hall, museum, and library are gathered together.
The waterfront landscape is presented again in this three-dimensional platform. The three cultural programs have greatly promoted the image of Culture Park, which adds to the charm of Changsha city. People may enjoy all cultural facilities any time in here at different elevated levels inside and outside of the building. Massive activities, be it exhibition and entertainment in the daytime, or carnivals in the evening, could be carried out here in this Cultural Park.

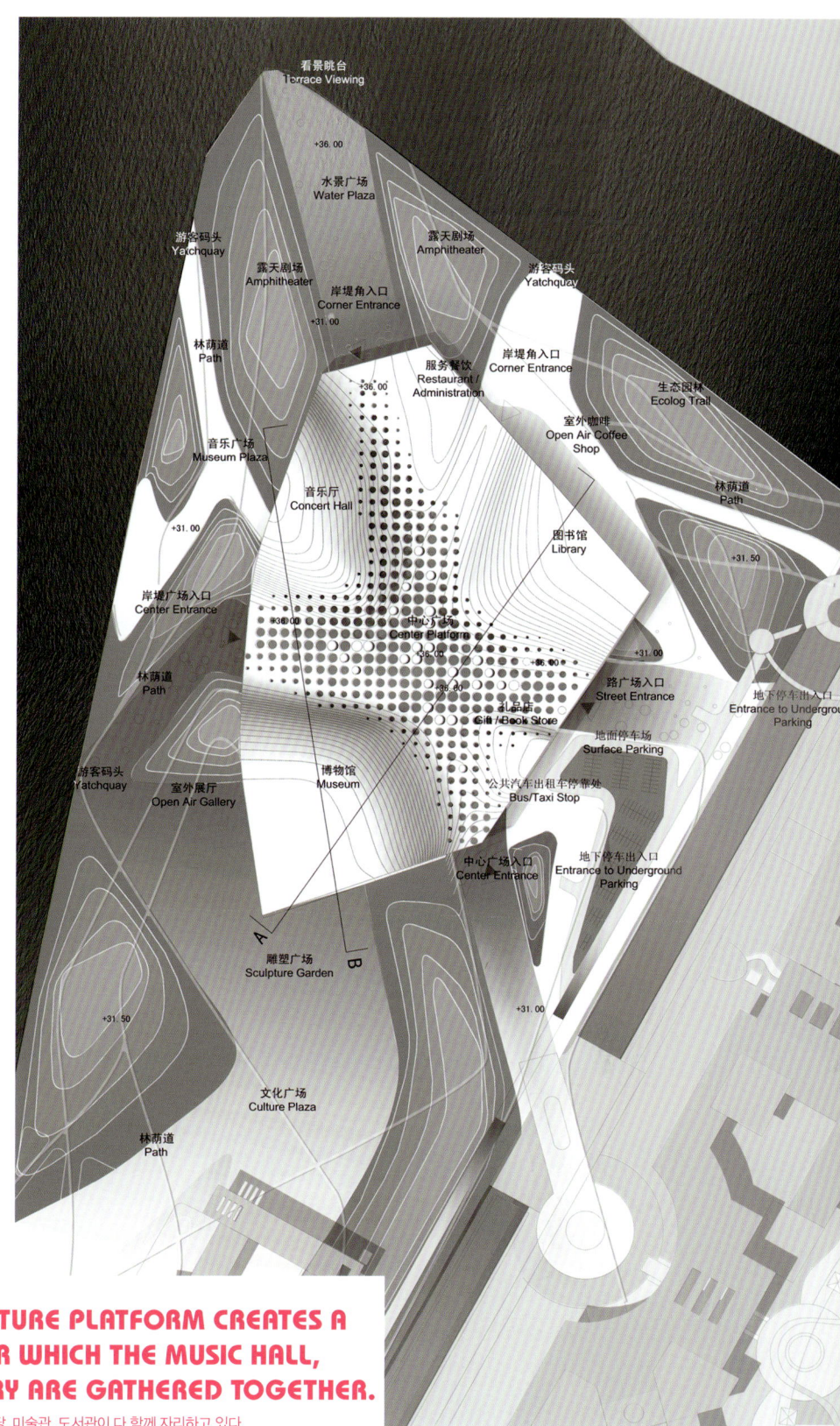

CHANGSHA NEW CULTURE PLATFORM CREATES A SOFT SURFACE, UNDER WHICH THE MUSIC HALL, MUSEUM, AND LIBRARY ARE GATHERED TOGETHER.
새로운 문화공간이 만들어 내는 곡면 아래 음악당, 미술관, 도서관이 다 함께 자리하고 있다.

Site plan

1F plan

ERDOS MUSEUM
에르도스 박물관

Inner Mongolia, China

Director in charge: Ma Yansong, Yosuke Hayano, Dang Qun **Design team:** Shang Li, Andrew C. Bryant, Howard Jiho Kim, Matthias Helmreich, Zheng Tao, Qin Lichao, Yang Lin, Sun Jieming, Yin Zhao, Du Zhijian, Yuan Zhongwei, Yuan Tao **Collaborator:** China Institute of Building Standard Design and Research, Institute of Shanxi Architectural Design and Research **Program:** Museum **Site area:** 27,760m² **Building area:** 41,227m² **Building height:** 40m **Client:** Erdos Municipal Government

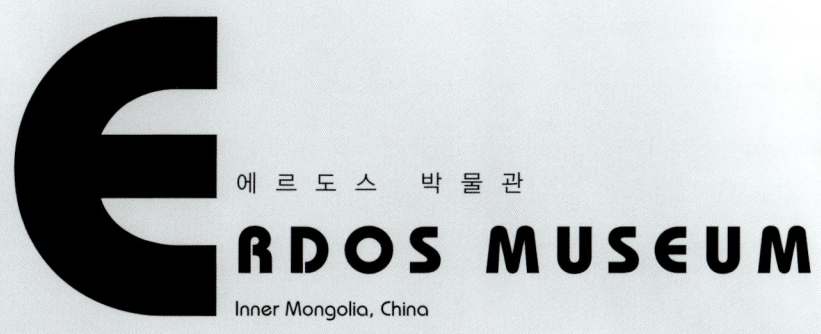

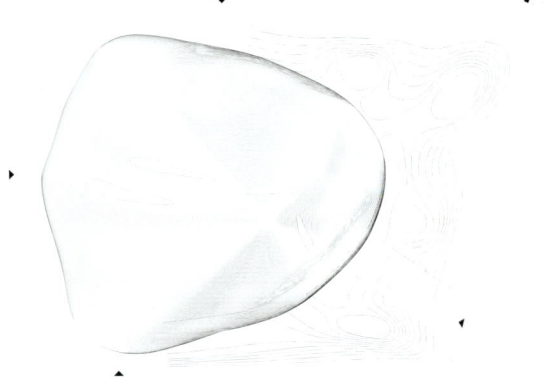

Master plan

에르도스 박물관은 현재 빠른 속도로 형성되고 있는 도시에 자리하고 있다. 활발한 경제 활동에 힘입은 에르도스 시정부는 기존 도시에서 수십 킬로미터 떨어진 곳에 새로운 시티센터를 세우기로 결심했다. 2005년까지 고비사막 외에는 아무것도 없던 이 지역에 '초원 위로 떠오르는 태양' 이라는 주제의 마스터플랜이 계획되었고, 계획안은 그곳에 살게 될 거주자보다는 정부를 만족시키는 아름답지만 의미 없는 이미지를 제안했다. 에르도스 박물관은 이 신도시 중심에 들어서게 될 것이며 우리는 마스터플랜을 고려해 그것에 반응하는 개념을 찾았다. 계획된 도시와 대조되는 자연스럽고 불규칙한 중심부를 설계해 건물 내부 풍경과 외부 풍경이 완전히 분리되도록 의도했다. 주변 풍경은 박물관을 감싸고 있는 금속의 반사성 루버를 통해 반영되고 조각난다. 새로운 공공장소로 떠오를 내부 공간은 연속적인 백색 곡면으로 연결된 여러 개의 전시실로 나뉘어 있으며, 유리로 된 지붕을 통해 빛을 유입하고 루버로 자연 통풍을 유도한다.

The Erdos Museum is located in a city that has recently been built at a rapid pace. Driven by a booming economy, the Municipal Government of Erdos, has been determined to build a new city centre, dozens of kilometers away from the current city. There was nothing but the Gobi Desert on this site in 2005. An urban masterplan was created, entitled 'Ever Rising Sun On The Grass Land.' This plan drew a beautiful, but empty image, one that fulfilled the wishes of the government, but doesn't hold much for the people who will have to live there. The Erdos Museum will be created at the centre of this new city. Our concept is a reflection and a reaction to the masterplan. The design is a natural, irregular nucleus, to contrast with the planned city, and to provide interior scenery completely separate from what is outside. The museum is wrapped in reflective metal louvers. The surrounding has been reflected and fragmented by the surface. The interior will become a new public space, divided into several exhibition halls, connected by continuous, white curvilinear walls. The glazed roof will let light into this space, while the louvers will allow for natural ventilation.

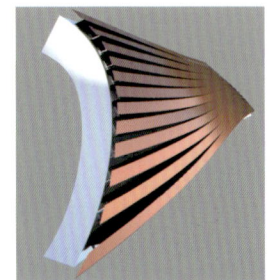 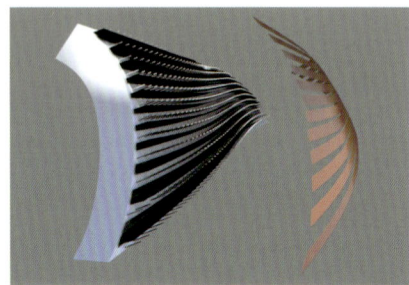

Louver

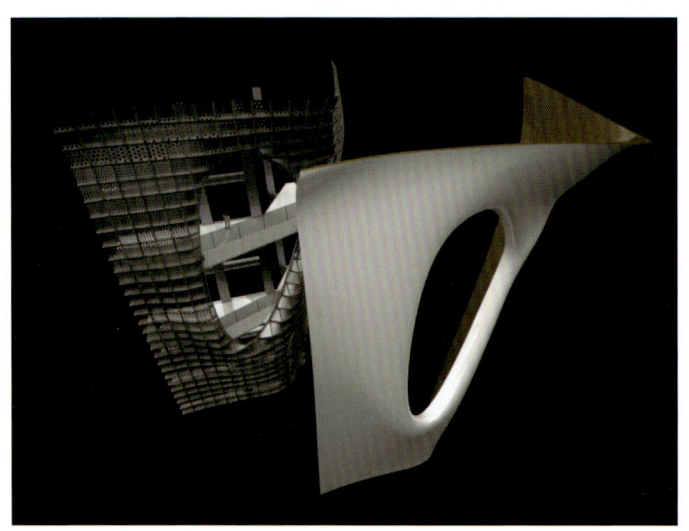

The design is a natural, irregular nucleus, to contrast with the planned city, and to provide interior scenery completely separate from what is outside.

계획된 도시와 대조되는 자연스럽고 불규칙한 중심부를 설계해 건물 내부 풍경과 외부 풍경이 완전히 분리되도록 의도했다.

Denmark Pavilion

덴마크 파빌리온

Denmark

Director: Ma Yansong, Yosuke Hayano, Dang Qun **Design team:** Yu Kui, Lie Jieran, Linda Stannieder, Louise Fiil **Program:** House
Building area: 269m² **Client:** Brain Stone

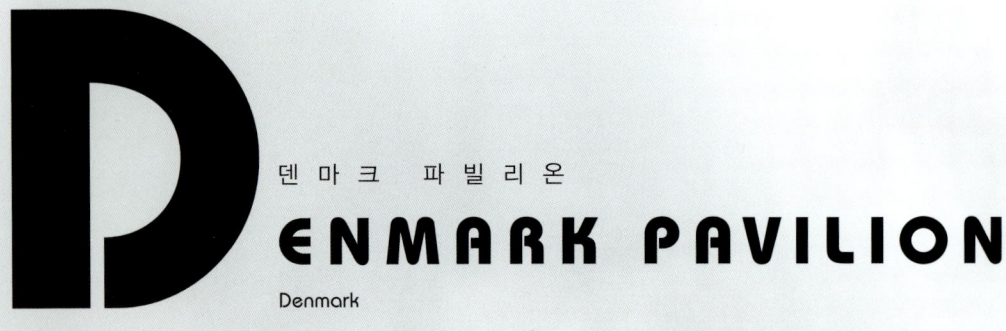

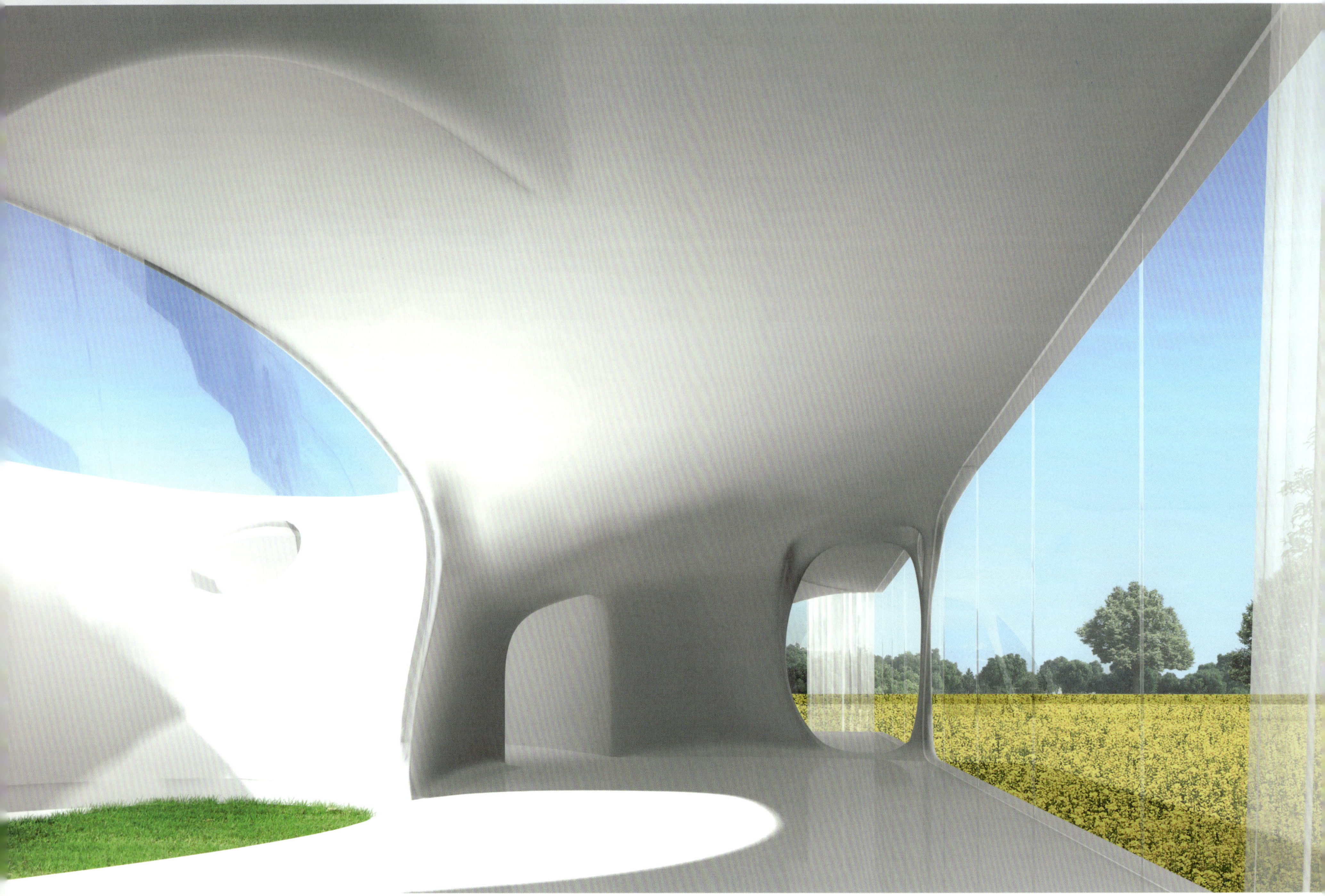

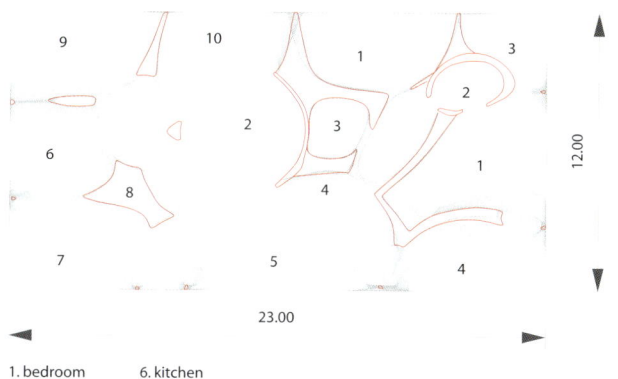

1. bedroom
2. courtyard
3. bath
4. library
5. living room
6. kitchen
7. dining room
8. mechanic
9. guest room
10. entrance

이 프로젝트는 삶과 구조가 공생하는 실험으로 2007년 9월 하나의 건물로 자라날 때까지 분열을 통해 100배로 성장할 태아다. 중국에서 시작한 여정은 코펜하겐, 바르셀로나, 베를린을 거치며 미스 반 데어 로에와 공간을 초월하는 대화를 나눈 뒤 덴마크의 작은 도시에 정착할 예정이다.
이는 유럽인들이 갖게 될 첫 '중국산' 주택인데, 일반적으로 생각하는 저가의 질 낮은 중국산 제품과는 달리 미래를 바꿀 수 있는 잠재력을 가진 유전자로 통제된, 생태적인 태생을 가진 진화의 결과다.

The project represents the symbiosis of life and structure; a part of an experiment, an embryo that is going to split and expand 100 times, until it grows into a building in September 2007. Then it will start its journey from China to Copenhagen, Barcelona, and Berlin, where it will have a trans-space conversation with Mies van der Rohe. After that, it will settle down in a small city in Denmark. The occasion will mark the first time that people in Europe will be able get a house wholly made in China. The house will be totally different from those low-priced, low-quality products associated with the "made in China" brand, because its origin is a life form; the outcome of an evolution controlled by a gene with the potential power to change the future.

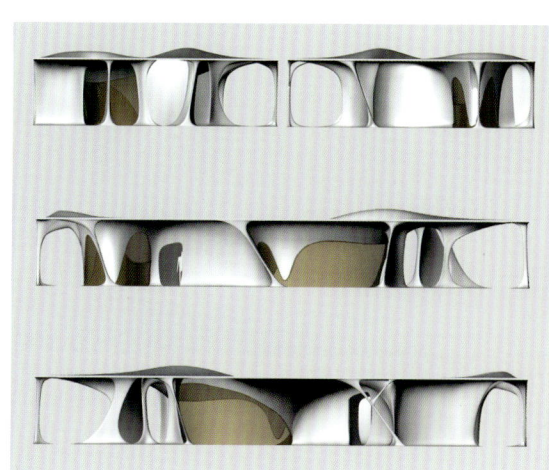

몽골리안 회원제 메도우 클럽
MONGOLIAN PRIVATE MEADOW CLUB
Inner Mongolia, China

Director: Ma Yansong, Yosuke Hayano, Dang Qun **Design team:** Guntis, Zhao Wei, Yu Kui, Evone Tam, Peng Li, Louise Fiil, Wang Xingfang, Fu Changrui, Wang Yuguo, Xu Yang, Liu Xiaopu **Program:** House **Building area:** 506,930m²

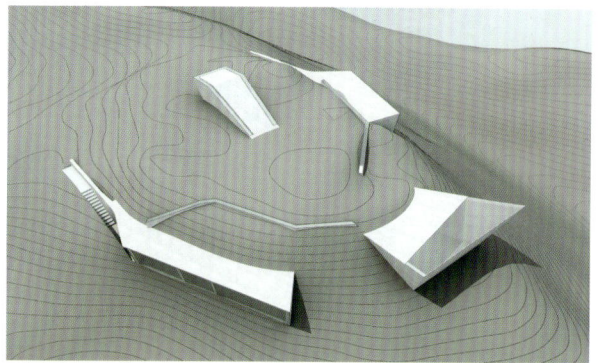

사람들은 오랫동안 도시의 삶을 지배해왔지만 예전에는 그 일부를 이루며 함께 생활했다. 내몽골 울란부통 초원 중심에 자리한 클럽으로 자연과 인간이 조화를 이루는 곳이다. 여름에는 잔디로, 겨울에는 하얀 눈으로 덮이는 500,000m²의 대지에서 보이는 모든 것은 자연적인 요소이며 건물은 그 안에 묻히도록 설계되었다. 건물의 주요 기능은 1년에 몇 주 동안 도시에서 방문하는 이들을 맞이하는 것으로 주변 풍경을 생활공간의 일부로 변환시켜야 하는 큰 과제가 주어졌다. 서비스 공간, 휴식 공간, 취침 공간 등 3개의 다른 레벨은 여러 개의 수직 계단으로 연결되며 서로 다른 기능 공간을 구분하고 있다. 건물 위치는 초원에 자연적인 요소를 더하는 기존의 흰 자작나무를 고려해 주의 깊게 선정되었다. 몽골리안 회원제 메도우 클럽은 공간이 자연이 되고 자연이 공간이 되는, 자연과 함께하는 생활공간이라는 개념 하에 자연과의 거리 개념을 새로 정리하고 있다.

Human being has been dominating the life of metropolitan for long time but people used to be part of it even living with it. Located in the middle of an extensive meadow in the Ulanbutong District of Inner Mongolia, the project aims to create a space that can connect humans and nature once more. With a 500,000sqm site area covered with grass in the summertime and snow in the wintertime, anything visible is natural matter and the building is designed to be buried in it. The main function of the building, for only few weeks a year, is to welcome people who come here from the cities, therefore the main design challenge is to convert the surrounding landscape and make it part of the living space. The three different levels of the building are: a service area, a relaxing area, and a sleeping area. All are connected with various vertical steps, so as to differentiate space for each functional zone. The location of the building was carefully selected according to existing white birch trees, whose white color express a fresh breath of nature in the middle of the grass field. The Mongolian Private Meadow Club shows the possibility to change the concept of distance within nature for creating living space. Space becomes part of nature, and nature becomes part of living space.

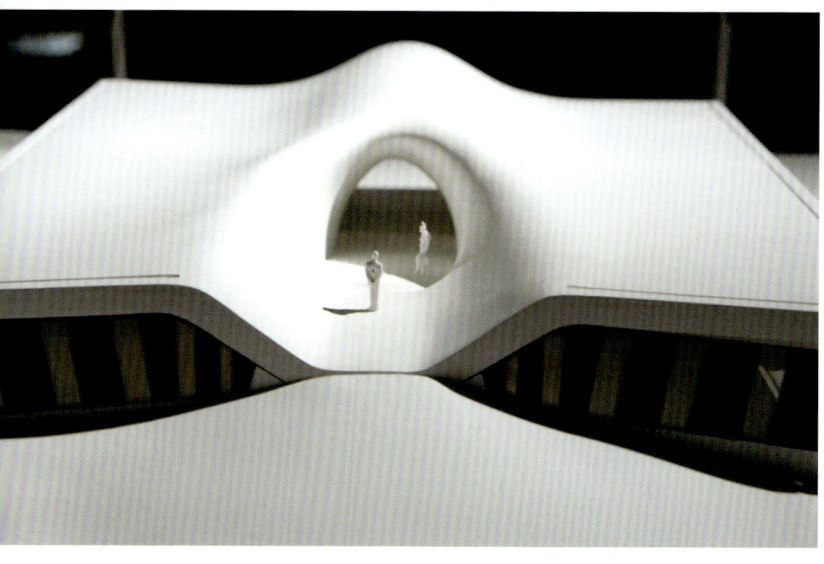

GUANGZHOU CLUBHOUSE

광저우 클럽하우스

Guangzhou, China

Director in charge: Ma Yansong, Dang Qun **Design team:** Yu Kui, He Wei, Stefanie Helga Paul, Erik Amir, Jia Huichao **Program:** Clubhouse **Site area:** 3,321m² **Building area:** 13,554m² **Building height:** 17m **Client:** Guangdong Xhongli Investment Co., Ltd.

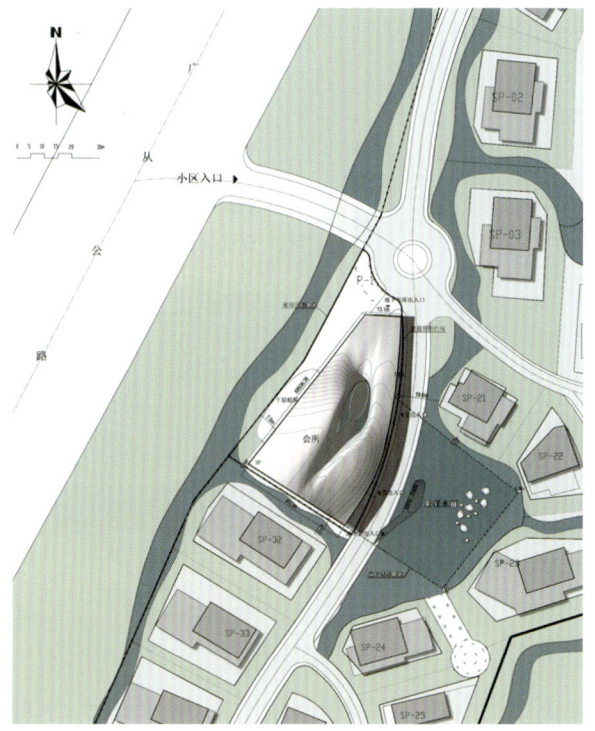

Site plan

1F plan

1. health station
2. cafe
3. beauty salons
4. atrium
5. post office
6. mechanic
7. neighborhood committee center
8. elderly center
9. gym
10. swimming pool
11. garage
12. toilet
13. restaurant
14. office

Section

중국 남부 도시 광저우에 자리한 이 클럽하우스는 주택 개발지 중심에 있는 시설로 주민들을 위한 공동시설을 갖추었다. 전체적인 시각으로 디자인에 접근했으며, 클럽하우스는 흐르는 듯 떠 있는 지붕 아래 있다. 양옆의 수면은 클럽하우스의 지붕을 반사하고 있으며, 이 모든 요소가 클럽하우스를 형성한다. 여기서 지면 위에 떠 있는 듯한 모습의 지붕이 중요한 역할을 하는데, 흐르는 형태로 다양한 높이를 만들며 지면과 다양한 관계를 끌어내고 있다. 이는 클럽하우스의 다양한 기능을 반영한다. 낮은 곳은 공공 로비 공간이며 4층 높이의 높은 곳에는 2개의 식당과 작은 호텔, 편의점 등이 있다. 지붕 중앙의 오프닝은 건물을 관통하며 지하층까지 빛을 유입한다. 클럽하우스 주변의 호수 또한 여러 요소를 담고 있다. 방문객을 위한 고요한 풍경을 연출하는 한편 수중 원형 오프닝을 통해 호수 아래 있는 실내수영장, 즉 물 아래 물로 빛을 유도한다.

The clubhouse is the hub of a residential development: a place to provide shared facilities for residents. This clubhouse will be located in the southern Chinese city of Guangzhou. This is a holistic design. The clubhouse is created beneath a fluid, floating roof. This roof is mirrored in the water that sits side-by-side with the structure. The clubhouse was conceived as a whole. The roof is key. It appears to float above the ground. The flowing shape creates different heights and different relationships with the ground. These reflect the diverse functions of the clubhouse: the lower end is used as a public lobby space, while the taller end, at four storeys high, can accommodate two restaurants, a small hotel and a convenience store. In the middle of the roof is a hole which drops down through the building, providing light to the basement levels. The water beside the clubhouse is also multi-faceted. As well as providing a reflective landscape for visitors, circular openings set in the water provide light to the area below. This area is an indoor swimming pool, set below the lake: water beneath water.

AL ROSTAMINI HEADQUARTERS
Dubai, UAE

Director in charge: Ma Yansong, Dang Qun **Program:** Office headquarters **Site area:** 4,392m² **Building area:** 50,000m² **Building height:** 80m

알 로스타미니 그룹 본사 건물은 두바이의 긴 해안을 따라 자리한 상업 중심지 내에 있다. 대중으로부터 해안을 차단하는 일반적인 커다란 오피스 블록 대신 우리는 그 위에 무언가를 짓고 싶었다. 우리는 해안 위에 떠 있는 얇은 슬라브 개념으로 출발해 모든 사무실에서 바다를 바라볼 수 있게 하는 한편, 지상층을 열어두어 대중과 해안이 직접 만날 수 있도록 계획했다. 슬라브를 지탱하고 있는 9개의 사선 튜브는 구조 안의 보이드다. 이 튜브들은 건물 후면을 관통하고 있으며, 그중 4개는 건물의 전면을 관통하며 마치 파사드에 구멍이 난 듯한 모습을 연출한다. 이들은 각기 다른 위치에서 각 층을 통과해 본사 건물이 가진 다양한 용도에 반응하는 다양한 사이즈의 공간을 만든다. 이 튜브들은 동선 역할도 하는데, 각각의 튜브는 건물 사용자를 각기 다른 층으로 이동시킨다. 튜브와 지면이 만나는 부분에는 싱그러운 식물들이 숲을 이루고 있어 해안의 경직된 풍경과 반사하는 수면 풍경에 나무들이 연출하는 부드러운 풍경을 더한다.

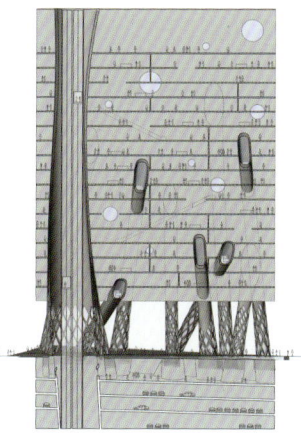 

Section

The site for the Al Rostamini Group headquarters is a long waterfront within the central business area of Dubai. Rather than creating a traditional big office block on this site and severing the water from the public, we wanted to build something above it. Our concept is a thin slab, which hovers above the waterfront. This gives every office a view over the water, and frees up the ground level, allowing the public direct access to enjoy the waterfront. The slab is lifted by nine diagonal tubes, voids within the structure. These tubes pierce the rear of the building, with four travelling through to appear as holes on the building's front facade. The tubes travel through floors in different locations, creating a variety of different sized, undulating spaces within the building. These spaces correspond to the diverse uses of the headquarters office building. The tubes also act as the method of circulation within the building: individual tubes transport the user to different floors. A lush forest is planted where these supporting tubes touch the floor. Thus, a soft landscape of trees gives way to the hard landscape of the waterfront, and the reflective landscape of the water itself.

BEIJING 2050

베이징 2050

Beijing Olympic Park, Beijing, China

2008 올림픽은 지난 수년간 베이징의 꿈과 야망을 상징했다. 하지만 우리는 2008년 이후 베이징의 미래를 상상하고 베이징의 장기적인 목표는 무엇이며 어떤 도시 디자인적인 새로운 가능성이 있는지를 고려해야 한다. 베이징과 정치는 뗄 수 없는 관계다. 중화인민공화국 10주년 기념 제10대 건축물과 아시안게임과 올림픽게임을 위해 지은 건축물을 포함해 베이징에 자리한 랜드마크적인 건축물 대부분은 현대사회를 향한 발전 과정에서 짧은 기간 안에 지어졌다. 이 건축물들이 도시의 모습을 바꿀 것인가? 여기서 도시는 도시의 이미지뿐 아니라 도시에 살고 있는 시민들의 삶을 포함한다.

우리는 시민들이 그들의 미래에 대해 생각하고 베이징을 꿈꾸며 더 큰 자부심을 갖길 바란다. 그것은 단순한 이미지가 아니다. 아름답거나 아니거나 그것은 역사와 오늘날의 세계를 돌아볼 수 있는 거울이며, 반항적이거나 급진적인 관점은 아니다. 우리는 역사와 현실을 직시해야 한다. 우리는 이 모든 것이 2050년에 이루어질 것이라 생각한다.

후통의 미래 _ 역사는 베이징의 귀중한 자산이다. 베이징의 찬란한 역사를 모르고 베이징을 이해할 수는 없다. 관광객에게 후통은 흥미로운 관광지지만 개별 화장실이나 샤워 시설이 없는 후통에 실제로 살고 있는 지역 주민에게는 매우 힘든 생활환경을 뜻한다. 후통 주민들은 정부에 의해 도시 외곽으로 강제 이주당하고 있으며, 한때 그들의 집이었던 곳은 부유층과 개발업자에 의해 새로운 장소로 탈바꿈하고 있다. 우리는 앞으로 자라날 세대들도 행복한 삶을 살길 원한다. 오래된 집은 새집으로 바뀔 것이고, 그것은 삶과 공간에 대한 자연의 법칙이다. 새로운 공간은 미래의 라이프스타일을 반영하고 옛것과 새것은 서로를 보완할 것이다. 2050년의 후통은 전통적인 모습뿐 아니라 사람들의 삶을 존중할 것이다.

중심업무지구 위로 떠 있는 섬 _ 베이징의 중심업무지구는 부와 지위를 상징하는 현대화에 대한 서양의 20세기형 비전에 따라 지어졌다. 하지만 베이징의 중심업무지구는 기술의 한계를 시도하려는 서양 국가들의 야망을 담고 있지 않으며, 미래를 위한 새로운 기준을 찾으려 하지도 않는다. 인구가 밀집된 중국의 미래 도시는 어떤 모습일까? 우리는 단절 또는 단순히 초고층을 고집하지 말고 직접적인 연결을 시도해야 한다. 중심업무지구 위로 디지털 스튜디오, 멀티미디어 비즈니스센터, 극장, 식당, 도서관, 관광지, 전시장, 체육관, 인공호수 등을 올리고 수평적으로 연결한다. 새로운 도시 구성 원칙과 더불어 이 제안은 근대주의의 '기계 미학'과 '수직 도시'에 대한 우리의 의문을 담고 있다.

천안문: 시민들의 공원 _ 오늘날 우리가 보고 있는 천안문 광장의 역사는 길지 않다. 지난 수십 년간 겪은 많은 변화는 진화하는 중국의 정신을 반영하고 있다. 2050년 중국은 더욱 성숙한 민주주의 국가로 탈바꿈해 붉은 광장처럼 대규모 정치 집회나 군대 행진을 위한 공간은 더 이상 필요하지 않을 것이다. 교통 시스템의 변화로 더 이상 지상 시스템에만 의존할 수 없어 공중 혹은 지하 시스템을 개발해야 할지도 모른다. 정치 혹은 교통 관련 기능을 상실한 천안문 광장은 어떤 모습일까? 지상은 정원이나 공원이 되고 지하에는 교통 네트워크와 연결된 문화시설이 들어설 수 있다. 공산당 시설들이 모인 현재의 중난하이의 형태를 흐트러뜨리는 '랜드스케이프 마운틴' 내부에 국립극장이 숨어 있을지도 모른다. 2050년 천안문 광장은 베이징 중심의 가장 큰 녹지와 삶이 가득한 도심 공간으로 변할 것이다.

The 2008 Olympic Games have symbolized the dreams and ambitions of Beijing for many years. However, as the 2008 Olympic Games drew to a close, it became necessary for us to envision Beijing's future beyond 2008, to think about Beijing's long term goals, and to imagine new possibilities for the city's design. Beijing and politics are inseparable. Almost all the landmark architecture in Beijing, including the "10th anniversary of PRC, 10 great architectures," projects built for the Asian Games and the Olympic Games, were built within a short period, at several stages of the modern society's development. Will these architectural creations reshape the city? The city we mean is not just comprised of its images, but also of the lives of the people living in it.

We hope it will lead to people thinking about their future and building greater confidence in their dreams in Beijing. This is not only about a picture, be it beautiful or not, it is perhaps about a mirror from which we can get a good look at history, as well as at the world today. It is not a rebellious or radical point of view, for we should be familiar with history and reality, and we believe all this will come into being in 2050.

The Future Of The Hutongs _ History is an invaluable asset of Beijing. The city can't be understood without a grasp of Beijing's rich history. While Hutongs are a haven for the visitors, they are part of a difficult living environment for local Beijingers, as these living spaces have no private bathrooms or showers. Furthermore, the Hutong dwellers are now being moved to outskirts of the city by the government, their homes being replaced by developers with homes for the wealthy. We hope that different generations can live happy lives in this land. Some new houses will replace old ones; that's the natural law of life and space. The new ones will reflect future life styles. The old and the new complement each other in their spaces, respectively. The 2050 Hutongs value people's lives, not just the traditional form.

Floating Island over the Central Business District _ The CBD in Beijing was built according to the Western vision of modernity created in the last century, which is currently regarded as an expression of wealth and status. However, Beijing's CBD isn't characterized by the Western ambition to push the limits of technology, nor does it attempt to set future new standards for itself. What will the densely populated, future cities of China look like? We think we need a literal connection rather than segregation, or simply chasing the building heights. Digital studios, multi-media business centres, theaters, restaurants, libraries, sightseeing, exhibitions, gyms, and even a man-made lake are elevated above the CBD, and horizontally connected with each other. This proposal and the new city organization principles articulate our queries of "machine aesthetics" and "vertical city," characterized by modernism.

Tiananmen: People's Park _ The Tiananmen Square we see today does not have a long history. All the changes it has witnessed in the past few decades reflect the evolution of the nation's spirit. By 2050, a mature and democratic China will emerge, and spaces for massive political gatherings and troop processions, such as Moscow's Red Square, may no longer be necessary. Public transportation will no longer be able to rely on the ground traffic system - it may utilize an above-ground or an underground system, due to changes in transportation patterns. What will Tiananmen Square be like when it will be deprived of its political and transportation functions? The ground might turn into a garden or park, and cultural facilities could be placed underground, to connect to the transportation network. A national theatre will be hidden inside a "landscape mountain," diffusing its forms in what is now Zhongnanhai, the nearby Communist Party compound. In 2050, Tiananmen Square will be an urban space filled with life, and the biggest green area in the centre of Beijing.

We think we need a literal connection rather than segregation, or simply chasing the building heights.
우리는 단절 또는 단순히 초고층을 고집하지 말고 직접적인 연결을 시도해야 한다.

SINOSTEEL INTERNATIONAL PLAZA
시노스틸 인터내셔널 플라자

Tianjin, China

Associate architects: Jiang Architects & Engineers **Program:** Office, Hotel **Building area:** 284,768m² **Height:** 320m **Engineers:** Ove Arup & Partners Hong Kong Ltd. **Completion:** 2012 **Client:** SINOSTEEL International Plaza (Tianjin) Ltd.

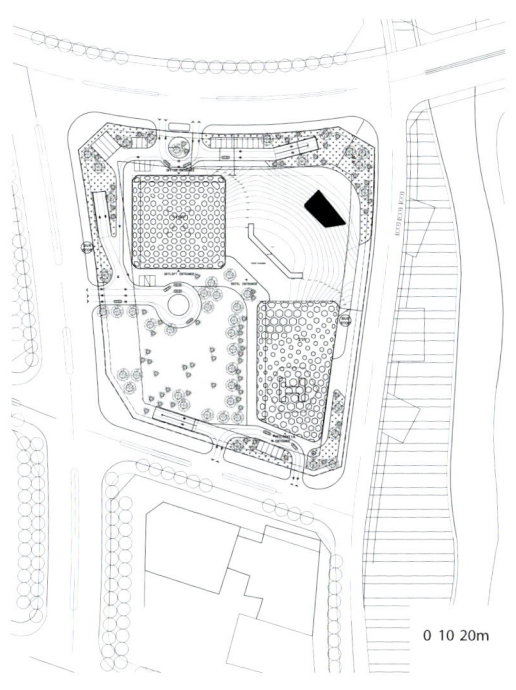
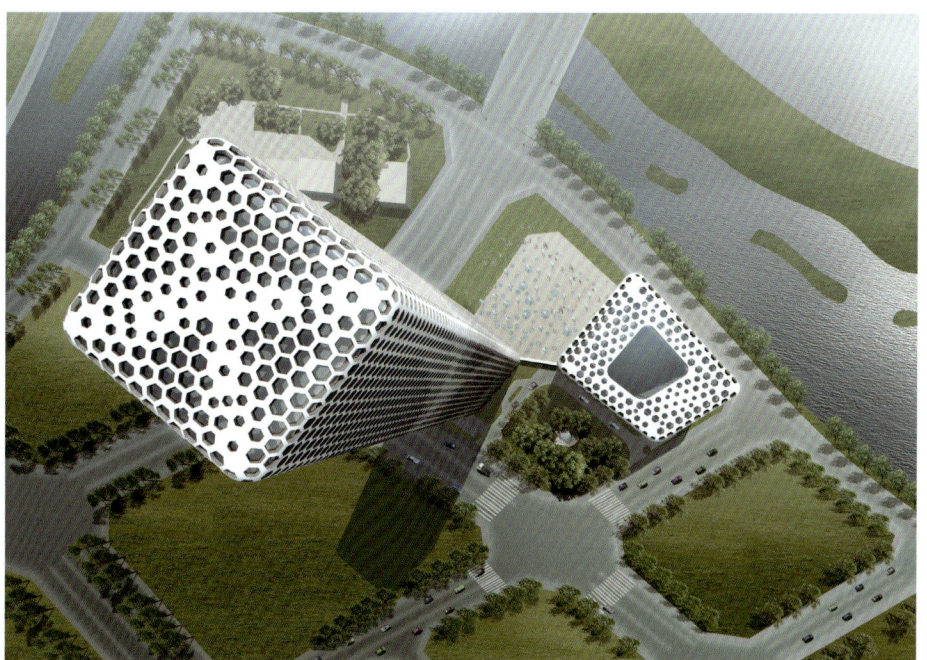

Master plan

중국 베이징의 동문 역할을 하는 텐진은 하이허강 어귀에 있는 수백 년의 역사를 가진 도시로 동양의 무한한 미래를 마주하고 있는 미래지향적인 시대의 정신이 살아 있다. 시노스틸 인터내셔널 플라자는 중국의 미래의 안정성과 위대함 그리고 시노스틸 인터내셔널 홀딩 사의 기업열과 정신력을 상징한다. 2012년 완공 예정인 이 건물은 높이 320m에 총 200,000㎡의 면적으로 이루어져 있으며 사무시설, 아파트, 호텔 및 기타 비즈니스 서비스 시설이 들어설 예정이다. 이 복합 상업시설은 진보적인 구조 설계, 에너지 및 환경 전략, 시공공법, 가상 환경 등 최신 건축 이론과 기술을 전통적인 방식과 접목시키며 역사에 대한 예의를 표한다. 육각형 창문은 중국 전통 건축과 정원에서 영감을 얻었으며, 이는 새로운 국제 사무시설을 정의하는 요소이자 계속해서 끊임없이 성장하고 번식하는 많은 세포로 이루어진 건물의 주 구조 시스템을 지지한다.

Tianjin is the East Gate of Beijing, China. The city is perched on the estuary of the Haihe River, and with its centuries-old history, it will face the boundless future of the East. Here, the spirit of the future and the feeling of the passing time gallop upward. The Sinosteel International Plaza building symbolizes the stability and mightiness of the Chinese future, as well as the enterprise spirit and vigor of the Sinosteel International Holding Company. The building, which should be completed by 2012, is 320 meters high, and has a net surface of about 200,000 square meters, destined for offices, apartments, a hotel, and related business services. This mixed-use centre expresses many of the most up-to-date theories and technical innovations in architecture, such as advanced structural design, energy and environmental strategies, construction and land engineering, and virtual environments, all used with regard and due respect for the traditional, and thus, history. Windows with six edges, in fact, are inspired from Chinese traditional architecture and gardens. They do not only become elements of new international office space design, but also shoulder the main structure system of this building, in order to make the building seem to be constituted of a lot of cells that grow unceasingly and reproduce continuously, in an endless succession.

1. pantry
2. printer center
3. storage
4. elec. room
5. air condition rooom

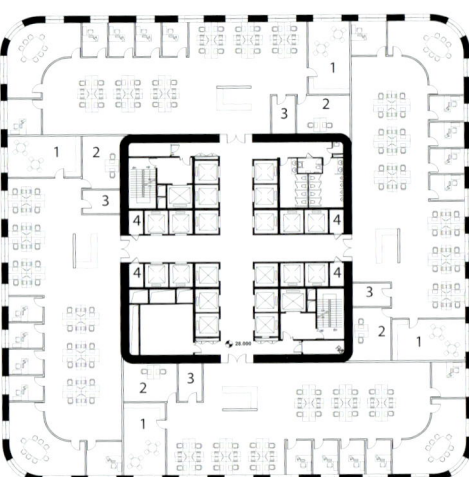

Office floor plan (5~45F)

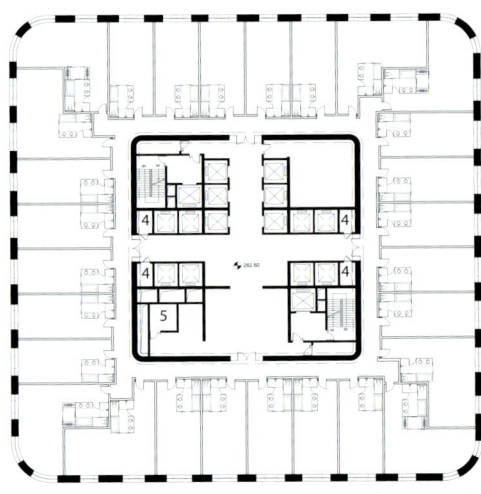

Hotel floor plan (61~69F)

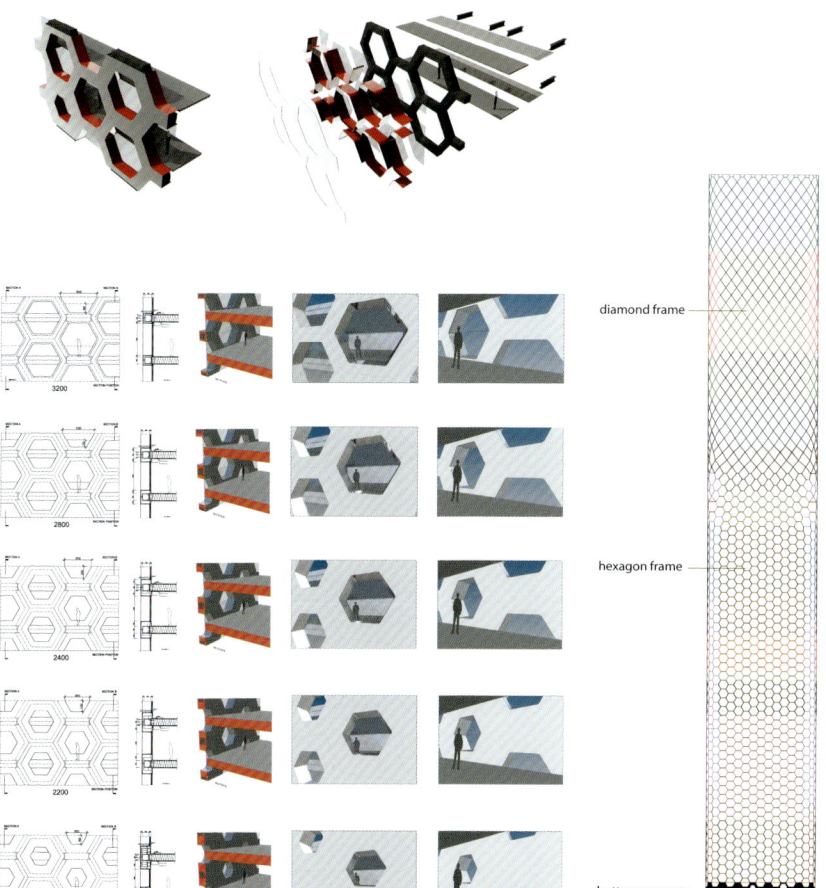

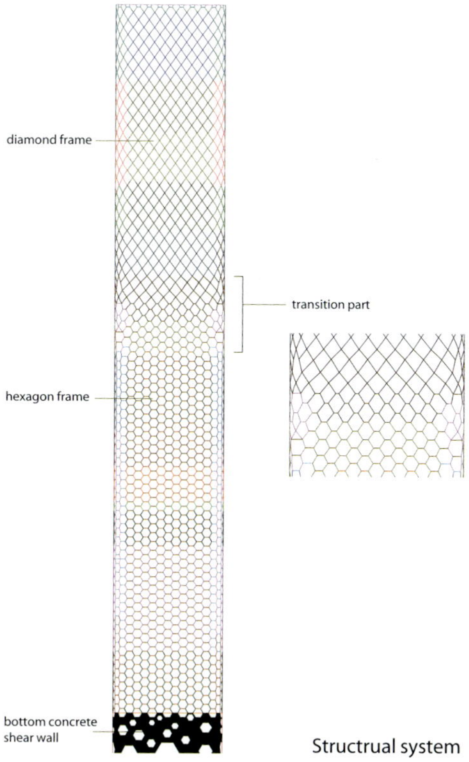

Structrual system

Water Tank for Goldfish

Designer: Ma Yansong, Yosuke Hayano **Prize:** Young Architects Award in New York, USA, 2006 **Size:** 300 (W) x 300 (D) x 400 (H) mm
Material: Acrylic

이 실험 '전기'에 우리는 물고기의 활동 궤도를 살펴보았다. 인간의 생활공간과 달리 물속에 사는 물고기의 세계는 중력 제한을 덜 받는 비교적 자유로운 환경이었다. 우리는 물고기의 활동 궤도에서 얻은 자료를 사용해 제한된 공간을 최대한 사용할 수 있는 디자인 전략을 만들었다. 형태 형성기인 디자인 과정의 '중기'에는 입체적인 공간에서 움직이는 궤도 분석을 통해 물고기의 움직임이 가장자리에서 빈번하게 일어나는 것을 발견했다. 그 결과 가장자리를 부드럽게 하는 것이 우리의 실험 목표 가운데 하나가 되었다. 프로젝트의 '후기'에 접어든 우리는 기본적인 벡터로 입력한 물고기의 운동 궤도를 공간 프레임으로 바꾸고 입방체의 공간을 최적화하기 위해 연속적으로 흐르는 공간 구성을 만들었다. 높은 빈도를 낮추기 위해 외부 표면을 변형하는 과정에서 다른 표면과의 접촉을 위해 안쪽으로 당겨졌다. 그 결과 외부 공간과 내부 공간이 역동적인 관계를 맺으면서 모호한 공간을 만들어낸다. 스테레오 리소그라피 모델링 방식을 적용해 물고기가 헤엄쳐 다닐 수 있는 유동적인 공간을 통해 건축적으로 혁신적인 조형성을 만들었다.

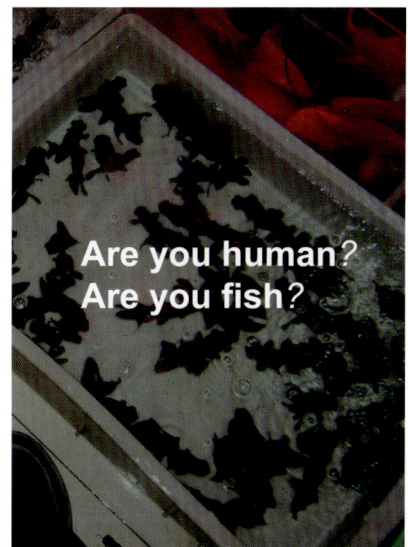

In the prophase of this experiment, we tracked the movement trajectories of fish. Different from human beings' living space, the water world is relatively freer of gravity restrictions. The fish's movement trajectory data became the initial driving force for our design strategy of maximizing, as well as optimizing the usage of limited space. At the metaphase of the morphogenesis design process, the analysis of the trajectories in 3-dimensional space showed a high frequency of fish's moving around the edges, hence smoothing edges became our experimental objective. In the anaphase, basic vectors of the fish's movement trajectory were transformed into an initial space frame, and a continuous fluid spatial organization was created to optimize cubical space. External surface was transfigured, in the effort to diffuse the high frequency of movement. In the meantime, the surface was strained inwardly to connect with other surfaces. Consequentially, the internal and the external space are dynamically related, and create an ambiguous space. During the phase, the stereo lithography modeling method was employed to allow the fish to circulate in fluid space, displaying an innovative architectural form.

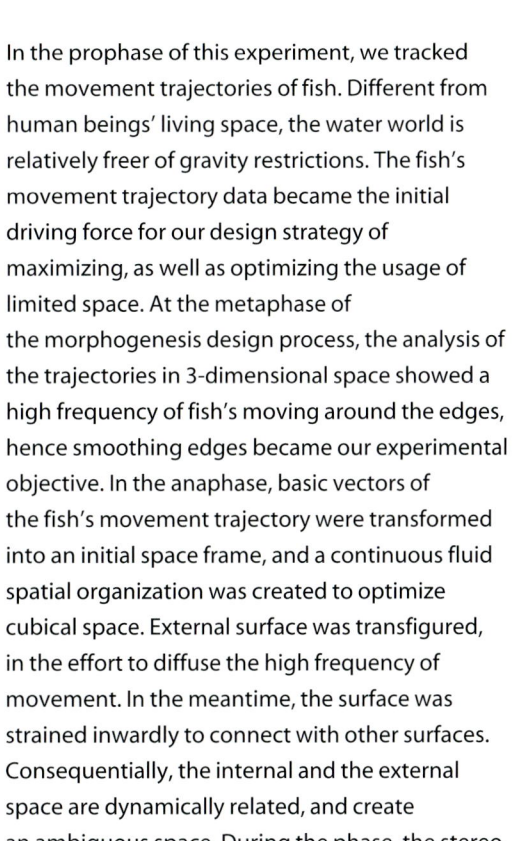

TOKYO ISLAND
도쿄 아일랜드

Dubai, UAE

Program: Resort **Height:** 20m **Building area:** 45,000m²

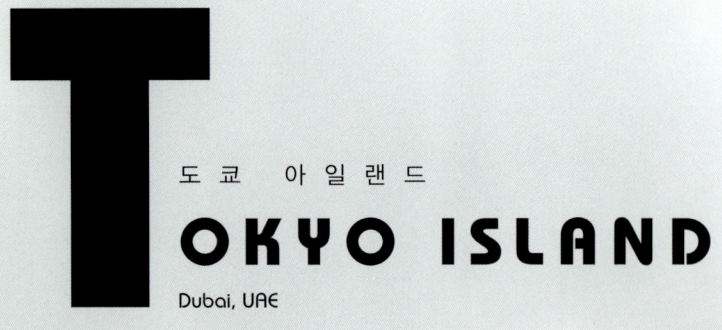

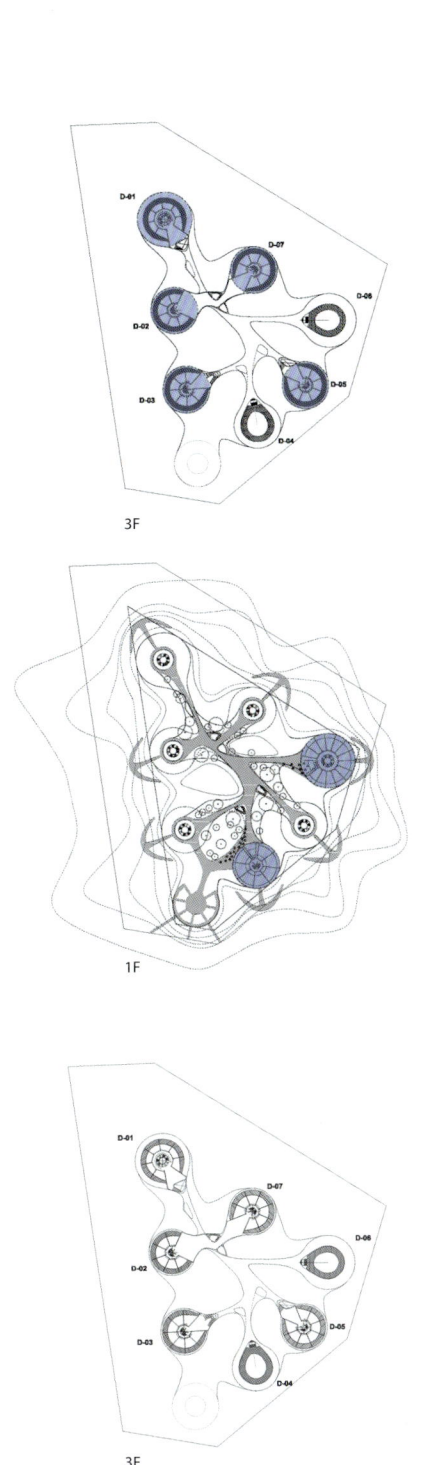
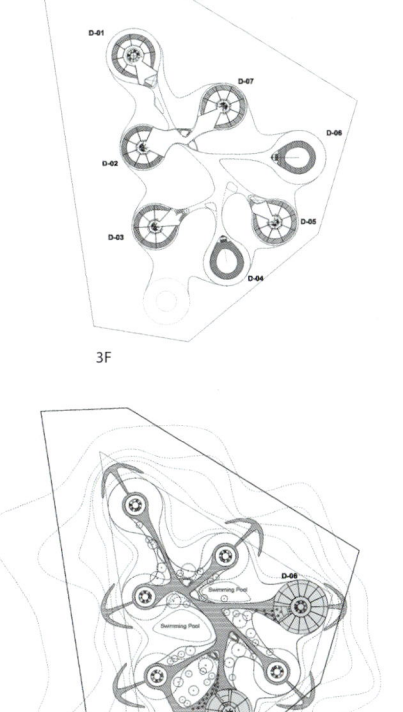

3F

4F

Residential density

1F

2F

3F

4F

1F

2F

Habitable floor

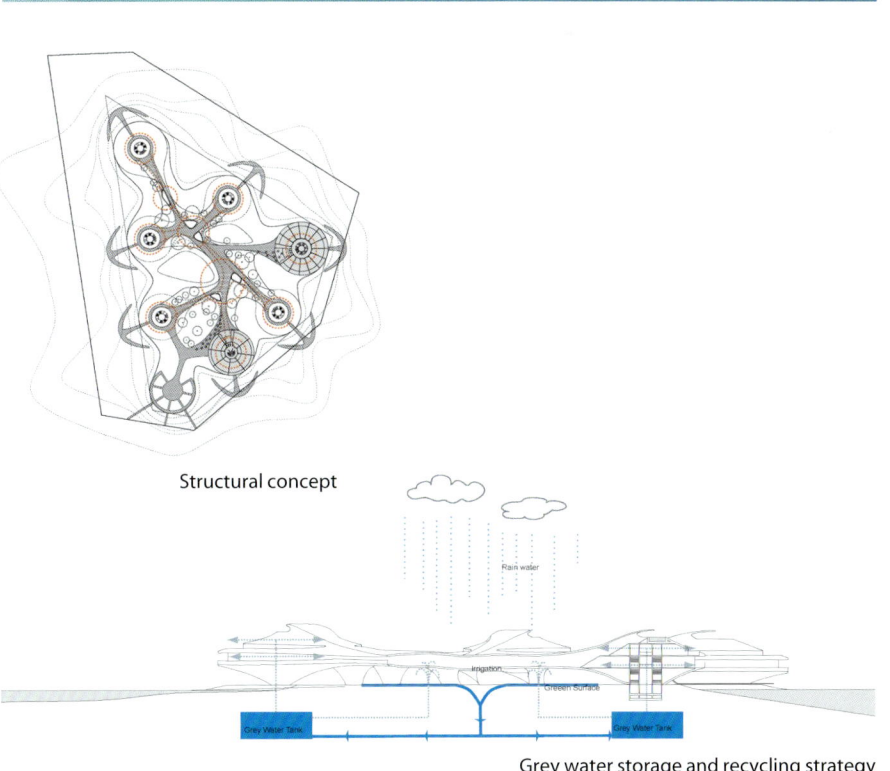

Structural concept

Grey water storage and recycling strategy

117

IAMEN MUSEUM
Xiamen, China

Director in charge: Ma Yansong, Dang Qun **Design team:** David Nightingale **Program:** Museum, Public space **Building area:** 13,340m²

중국 남부의 샤먼은 기후가 매우 따뜻하고 습하다. 새로운 시립미술관은 도시를 관통하는 호수 중앙에 자리한 섬이라는 영감적인 대지를 갖고 있다. 우리는 주변 풍경을 반사할 수 있는 자연적인 것을 만들고 싶었다. 우리는 땅 위에 떠 있는 미술관을 만들어 샤먼 시민들에게 커다란 오픈 스페이스를 주고 싶었다. 미술관은 3개의 층으로 나뉘어 있는데, 지상층에는 모두를 위한 오픈 스페이스가 있다. 방문객은 이 공공공간의 랜드스케이프를 통해 극장, 야외 운동경기장, 만남의 장소 등 다양한 여가 공간으로 여과된다. 버섯과 같이 자연스러운 형태의 중간층은 전시실, 카페, 식당, 사무실 등 미술관의 주요 기능을 담고 있다. 물과 잔디 등 랜드스케이프적인 요소로 이루어진, 대중에게 개방된 지붕에는 샤먼의 기후를 고려해서 설계한 태양 전지판이 설치되어 있다. 호수 중앙에 높이 자리한 미술관은 도시를 향해 확 트인 시야를 보장하고 건물 외관은 도시의 이미지를 반사하고 있어 호수 어느 곳에서나 미술관을 잘 볼 수 있다. 미술관은 샤먼의 새로운 랜드마크로서 샤먼에 많은 관심과 활기, 방문객을 끌어올 것으로 기대된다.

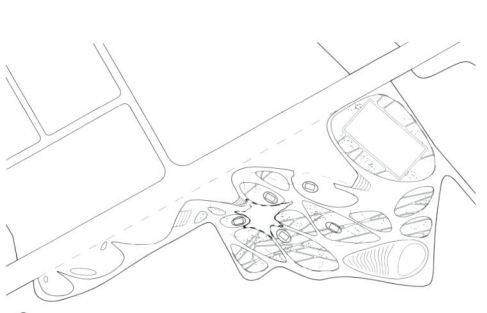

Section

Xiamen is a city in Southern China, with a very warm, humid climate. A new city museum will be created here. The site of the new museum is very inspiring: an island in the middle of a lake that runs through the centre of the city. We wanted to create something natural to reflect this landscape. We aimed to create a large, open public space for the people of Xiamen, a floating museum above the ground. The museum is split into three layers. At ground level, there is open public space. The landscaping of the public space filters people around the site to various recreational spaces: theatres, open sports fields, places to meet. The middle level is the building, a natural, mushroom-like shape. This level is devoted to the museum's functions: exhibition space, cafes, restaurants and offices. On the roof, we return once more to landscape: there is water, grass, public access. The roof is also home to solar panels, to take advantage of Xiamen's climate. The museum's position, high above the centre of a lake, ensures uninterrupted views of the city beyond. At the same time, the building's mirrored exterior reflects images of the city, making it visible from all angles around the lake. The museum will become a landmark, helping to attract attention, regeneration, and visitors to Xiamen.

Site plan

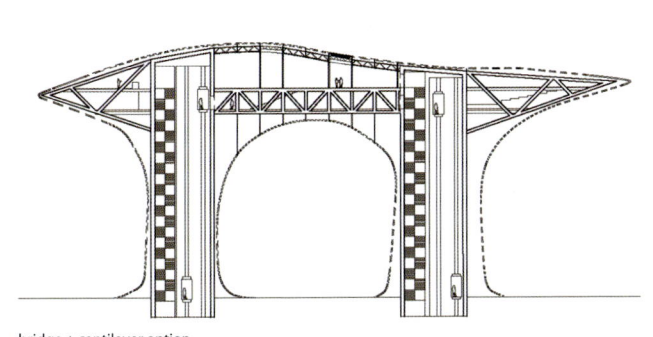

bridge + cantilever option

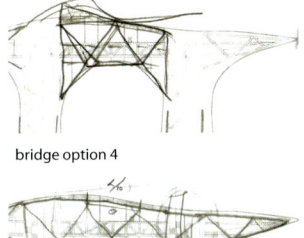

bridge option 4

cage option

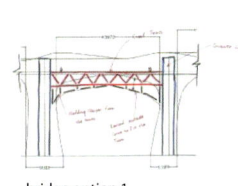

bridge option 1

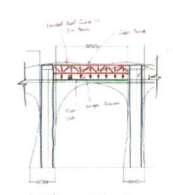

bridge option 2

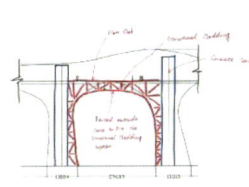

bridge option 3

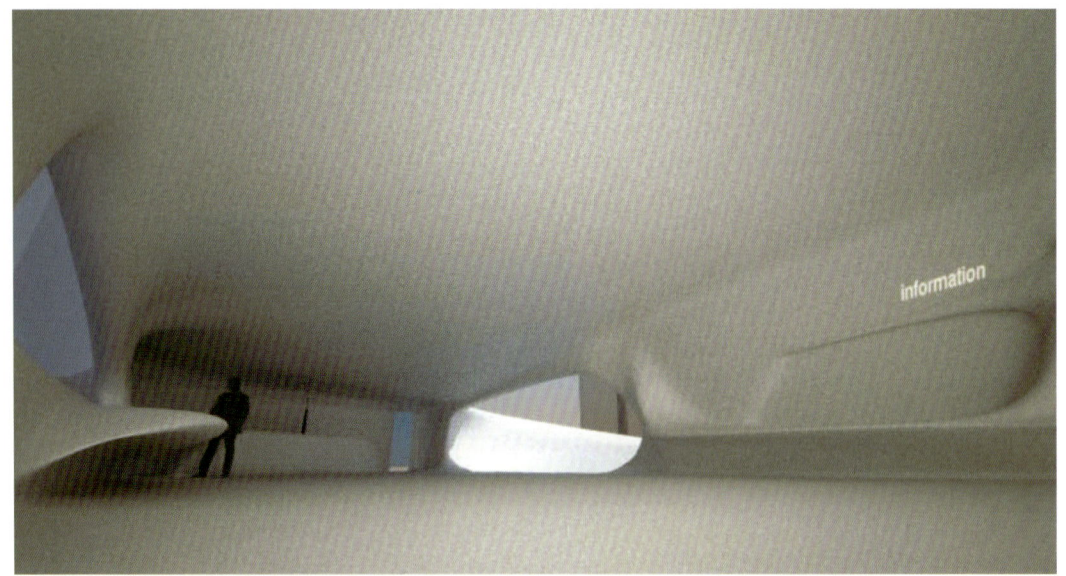

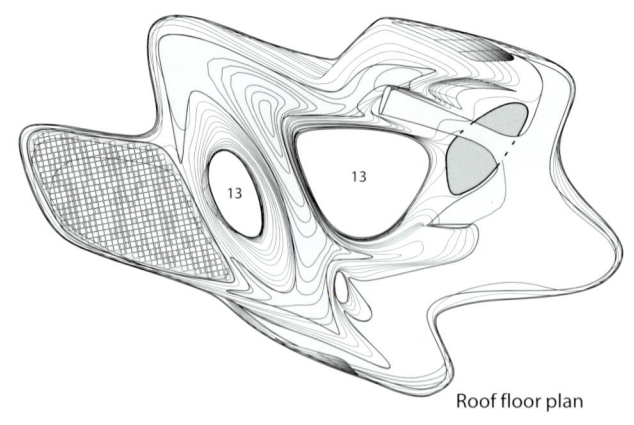

Roof floor plan

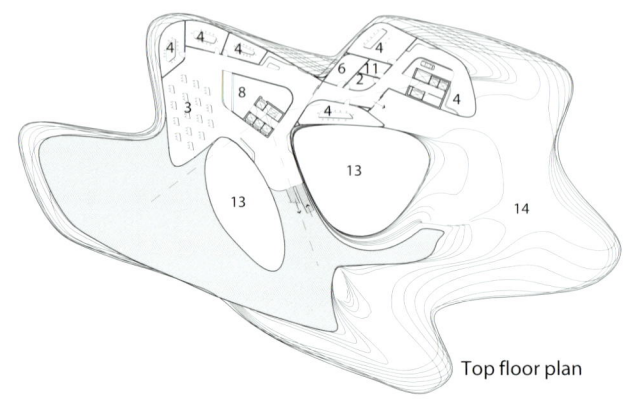

Top floor plan

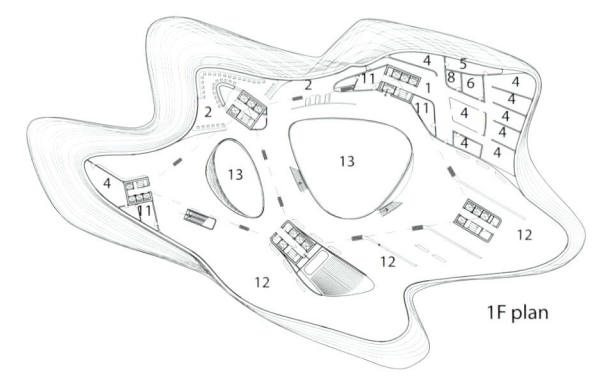

1F plan

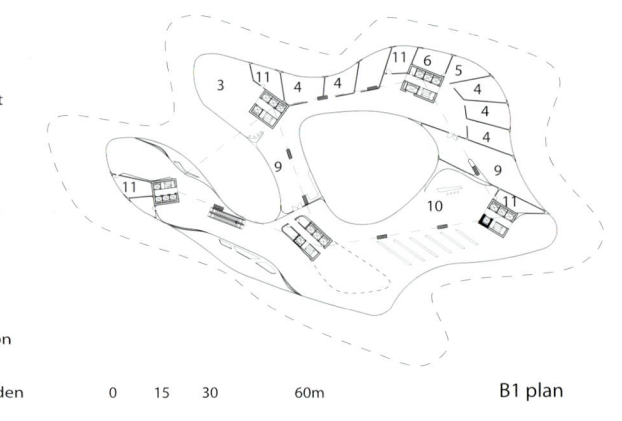

1. lobby
2. cafe
3. restaurant
4. office
5. restroom
6. storage
7. mechanic
8. kitchen
9. lounge
10. library
11. toilet
12. exhibition
13. void
14. roof garden

0   15   30   60m

B1 plan

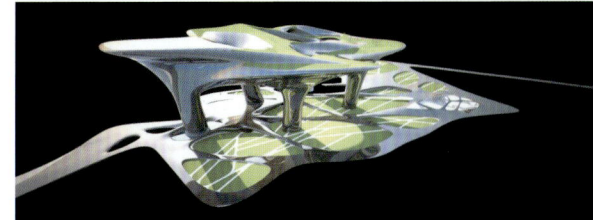
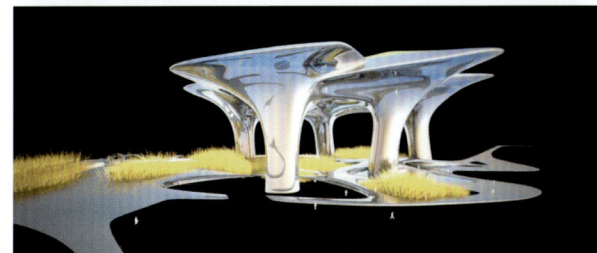

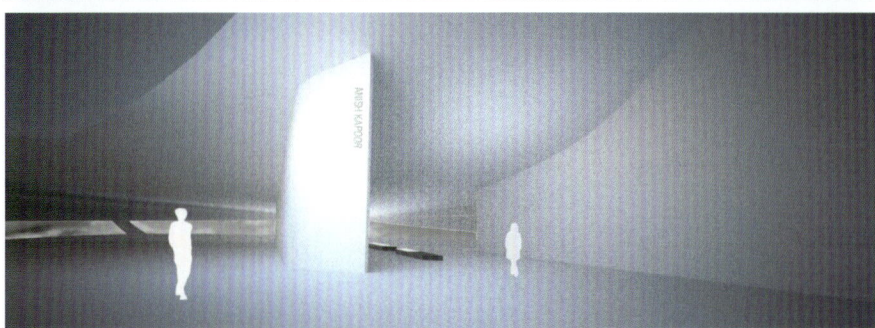
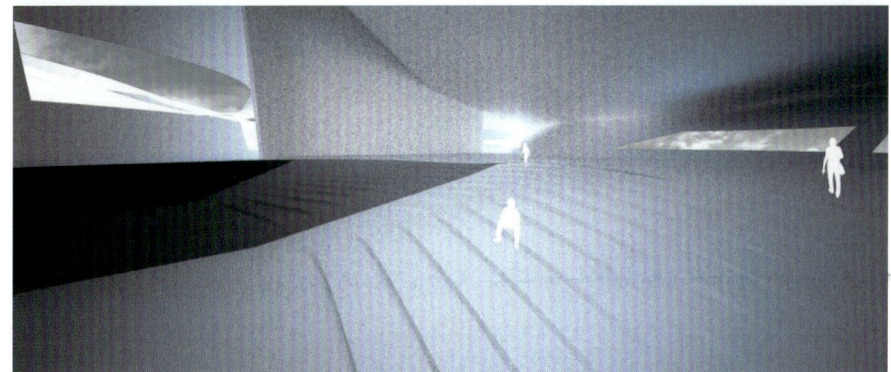

1. lobby
2. cafe
3. restaurant
4. office
5. restroom
6. storage
7. mechanic
8. kitchen
9. lounge
10. library
11. toilet
12. exhibition
13. void
14. roof garden

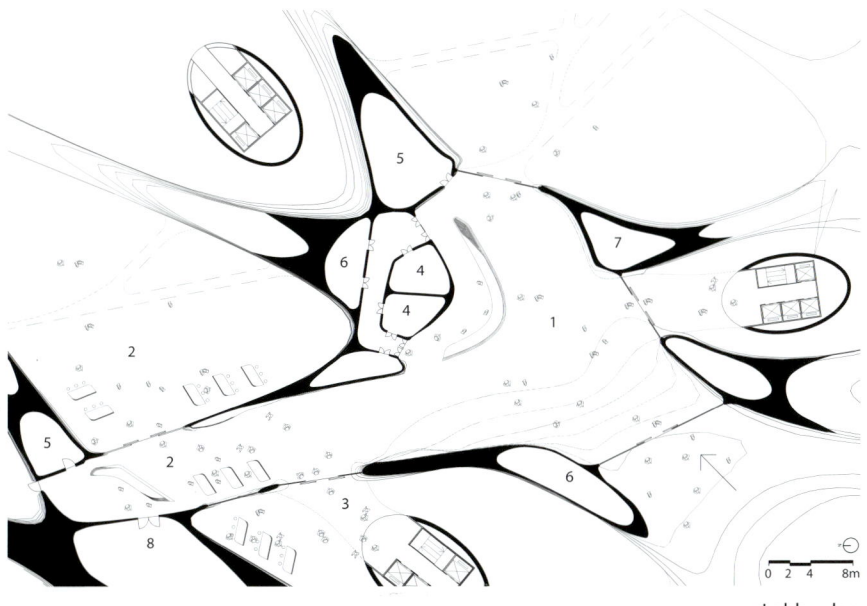

Lobby plan

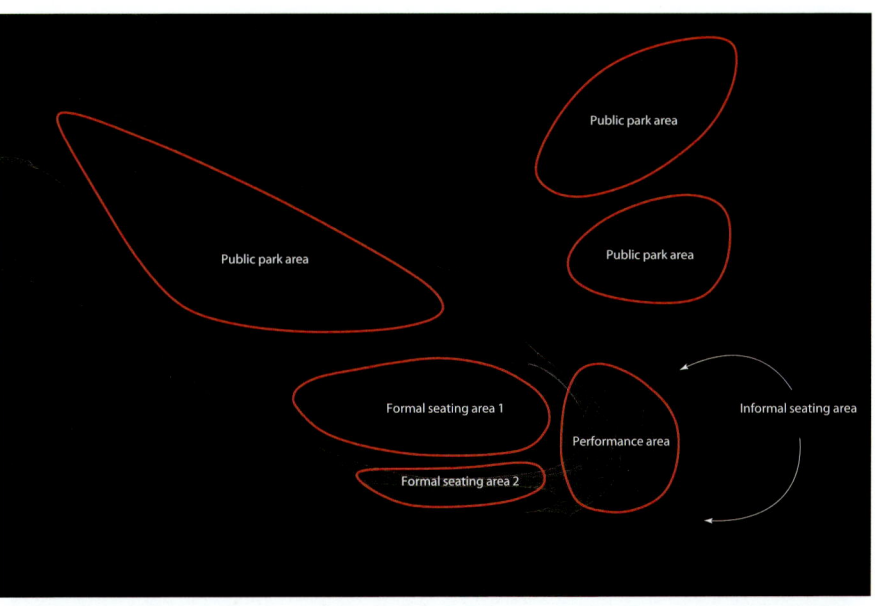

amphitheatre: diagram

structure: lighting/ sound system/ installations

formal seating area   performance area   informal seating area

WHY MAD IS MAD
왜 MAD는 열광적인가

Jiang Jun | 지앙 쥔

2004년 4월 1일 만우절, 해외에서 공부를 마친 마 옌송은 2년 전 뉴욕에 설립한 그의 건축사무소 MAD를 중국에 소개하기 위한 공개 기념식을 열기로 결심했다. 일반적으로 건축사무소 이름은 설립자의 사상이나 신념을 매니페스토적으로 표현하고 있다. MAD는 '마 디자인(MA-Design)'의 약어일 뿐이라는 마 옌송의 주장에도 불구하고 그 이름이 주는 언어적 암시는 MAD 특유의 시니시즘을 떠올리게 한다. 그리고 불과 3년 만에 그 이름과 마 옌송의 얼굴은 최신 유행하는 여러 잡지의 커버를 장식했다. 중국 신세대 건축가를 대표하는 그의 작품은 할리우드 히트작만큼이나 우리를 흥분시키며 그의 성공 비결을 '마 옌송 법칙'이라고 부른다. 무엇으로 MAD의 갑작스런 유명세와 스타덤을 설명할 수 있는가? 단순한 사무실 이름에서 인기 브랜드로 탈바꿈한 원동력은 어디에서 나오는가? 왜 MAD는 열광적인가?

중국 건축가와 그 세대가 겪은 사회적 배경을 이해하기 위해서는 '마 옌송 법칙'을 타이밍, 장소, 사회적 네트워크의 요소로 나누어봐야 한다. 첫째로 타이밍은 국제 건축사회에서 중국 건축가가 갖는 장점, 즉 그의 중국다움이다. MAD는 중국이 보다 선진화되고 현대화된 국가로 도약하려는 중대한 시기에 설립되었다. 공간적인 측면에서 현대화는 대규모로 이루어지는 빠른 도시화를 의미하며 이는 곧 건설 산업과 많은 건축가에게 활발한 시장을 형성해준다.

한편 문호 개방 정책을 이은 대약진 정책이 저물면서 강한 문화적 자의식을 가진 세대가 나타나기 시작했고, 건축계에는 창의적인 건축가들이 출현했다. 연간 수백억 위안 규모의 시장, 고가의 디자인 비용, 중국의 도시화에 참여하며 얻는 개인적인 성취감 등으로 건축가는 유명세와 높은 소득 모두를 얻을 수 있는 이상적인 직업으로 떠올랐다. '스타 건축가 시대'의 출현은 시대정신을 반영하고 있다. 동시에 세계는 중국을 주목하며 빠르게 변화하는 중기 왕국에서 어떤 '스타' 들이 탄생할지 궁금해했다. 어떤 시각과 신념을 가지고 세계적인 무대 속 중국을 이끌지, 혹은 그들은 중국에 어떤 변화를 가져올지 등을 호기심 어린 시선으로 주목했다.

'중국다움'만큼 중요한 요소로 '국제적임'을 들 수 있다. 미국 명문대를 졸업하고 뉴욕에 등록된 사무실 자격증, 세계 최고의 설계사무소에서 쌓은 인턴 경험을 갖춘 마 옌송은 그와 비슷한 국제적인 배경을 가진 파트너들과 함께 여러 국제공모전과 전시에 적극적으로 참여했다. 이러한 전략의 결과 2006년 MAD는 '마릴린 먼로 빌딩'으로 더 잘 알려진 토론토의 앱솔루트 타워 설계경기에 당선되며 일약 스타로 떠오르기 시작했다. 중국으로 돌아가기로 마음먹은 마 옌송의 결정은 국제무대에서 중국의 존재를 강화하고자 외국의 전문 인력을 수입하는, 중국의 '들여오고 내보내기'라는 전략과 잘 맞았다.

같은 시기 비슷한 해외 교육 배경을 가진 두 명의 선배 건축가 장융허와 마칭윈은 두 곳의 유명한 미국 대학 내 건축학교를 이끌게 되었지만, 먼로 빌딩은 주요 국제공모전을 통해 중국 건축가가 해외로 비즈니스를 확장한 첫 경우였다. 민족주의적 콤플렉스를 가진 국내 언론은 야오밍의 NBA 진출이나 리 우시앙의 세계기록 수립처럼 국가의 수준을 한층 끌어올리는 스토리로 다루었다. 현대화가 한창 진행 중인 중국은 국가의 성장과 국가적 르네상스의 지표가 될 기념비적인 이벤트를 필요로 한다.

중국다움은 세계의 이목을 끌며, 국제적임은 중국을 흥분시킨다. 안에 있는 사람들은 나가고 싶어 하고 반대로 밖에 있는 사람들은 들어오고 싶어 하는 '포위된 요새'의 논리는 중국의 근대성이 아직 부족함을 잘 나타낸다. 또한 그러한 논리에 근거해 장융허, 마칭원, 마 옌송처럼 이중적 배경을 가진 건축가들이 국내외 상황 모두에 대응할 수 있는 것이다. 그 상황은 자유로운 생각을 가진 차세대 중국 건축가들의 공통적인 입장이며, 마 옌송은 그들이 앞으로 해외에서 돌아올 학생들에게 큰 영향력을 미칠 것이라 주장한다. 마 옌송은 그와 같이 서양 문화를 접하고 돌아온 이들을 이끄는 대표적인 건축가가 되길 희망한다.

마 옌송은 담론할 권리를 얻기에 앞서 그 두 가지 전략의 이중적인 역할을 변증법적으로 이해해야 한다고 주장한다. 그는 어디에서 그런 권리를 얻겠다는 것이며, 무엇에 대해 담론할 것인가?

마 옌송의 시대에는 기본적으로 세 가지 종류의 건축 조직이 있다. 첫째는 국영 건축 디자인 하우스로 공공기관의 일부인 그들은 건설시장에서 큰 몫을 차지하며 중국 도시 풍경의 상당부분을 책임지고 있다. 둘째는 중국 건설시장의 개방과 지역 및 산업 규제 완화와 함께 밀려 들어온 국제 설계사무소들이다. 그들은 국제적인 수준을 유지하며 중국 도시의 건설 산업에 신선한 바람을 불러 일으켰다. 마지막으로 지역 중심의 개인 설계사무소가 있다. 1990년대 이후 하향식 계획에서 시장 위주 경쟁으로 도시 건설 구도가 바뀌며 개인 사무소들도 업계의 주요 구성원으로 떠올랐다.

셋 모두 제각기 특징을 가지고 있는데, 국영 디자인 하우스들은 주로 인맥을 기반으로 한, 수년간 쌓은 정부와의 경험을 자원 삼아 그들의 독점적인 위치를 상당 기간 유지할 수 있을 것으로 보인다. 이들은 주로 정부와 관련된 대규모 프로젝트 위주로 작업하며 관료주의 운영체계, 기계적인 가치관, GDP 중심의 빠른 생산 모드 등으로 보수적이고 진부한 성격을 면치 못한다. 하지만 사회·정치적 변화 속에서 권력을 상실한 몇몇 디자인 하우스들은 비교적 독립적인 스튜디오를 설립하며 그들만의 '스타 건축가'를 탄생시켰다.

국제적인 설계사무소는 기업형과 스타형으로 나뉘는데 대부분의 다국적 기업과 마찬가지로 기업형 설계사무소는 자금, 기술력, 행정력을 두루 갖추고 있다. 반대로 스타형 사무소는 진보적인 사고를 최우선시한다. 선진국에서 경험한 풍부한 경험을 바탕으로 한 뛰어난 접근 방식으로 프로젝트를 분석하며 경쟁력을 앞세우는 외국 설계사무소에 비해 국영 회사들의 경험은 부족하지만, 스타형 사무소는 중국의 현실을 꿰뚫지 못하고 지역 건축계의 이치를 이해하지 못하기 때문에 지역 보호 정책에 따라 움직이며 국영 설계사무소와 협력해 작업한다.

국영 디자인 하우스의 권력도 없으며 국제적인 설계사무소의 명성도 가지지 못한 개인 사무소들은 가장 소극적인 상황에 놓여 있으나 한편으론 양자의 장점을 모두 이용하기도 한다. 개인 설계사무소의 설립자들은 대부분 국영 하우스 또는 해외 사무소에서의 경험을 가지고 있으며, 중국의 현실을 이해할 뿐 아니라 국제적인 상황도 잘 알고 있다. 그들은 국내 사업과 정부 차원에서 말없이 행해지는 관습을 잘 알고 있으며, 국제무대의 최신 개념도 작업에 적용시킬 만큼 뛰어나다. 비록 이들이 건축시장에서 차지하는 비율은 다른 두 종류보다 낮지만 양자 모두를 활용할 수 있다는 것은 진정으로 영향력 있는, 이론과 실무를 겸비한 스타 건축가를 만들어낸다. 외국의 스타형 사무소와 맞서 지역 영향력을 차지하고 있으며 이는 곧 머지않아 세 종류의 사무소들의 몫이 비슷해질 것이라 예상하게 한다.

담론적 권리에서 우리는 시장 점유율이 거꾸로 된 것을 알 수 있다. 중국의 비판적 건축 담론은 아직 미숙하며, 현재는 대부분이 우수한 개인 사무소를 중심으로 행해지고 있다. 최근 중국과 해외에서 열린 중국을 주제로 한 주요 전시들과 대범한 아이디어를 담은 대화록들이 출판되면서 그 수는 적지만 많은 주목을 받고 있는 중국의 스타 건축가들이 형성되고 있다.

중국 건축계 내 이론과 실무의 부조화는 웅장한 이상주의와 멀리 내다보지 못하는 실용주의의 대립, 즉 '실무에서 부족한 부분은 담론으로 채운다'로 해석할 수 있다. 인맥과 이윤에 집착하는 건설시장을 겨냥해 대중매체를 통해 광범위한 가치를 알리기 위해 건축계는 '선구자'와 '순교자'를 위한 몇 개의 기념비를 세워야 할 듯하다.

주택이 상업화되고 건설의 열기로 나라 전체가 뜨겁게 달아오르면서 중국은 엘리트주의적인 '신 건축 운동'에서 '신 스타 건축가 운동'으로 변해갔다. 이는 개념이 중시되는 베이징과 같은 도시에서 더욱 특징적으로 드러나며, 여기서 스타 건축가들은 함께 일할, 새로운 것을 간절히 희망하는 부동산 개발업자들을 쉽게 찾을 수 있었다. 스타 건축가의 이름과 얼굴은 빌보드와 계약서를 장식하며, 그를 향한 갈채의 말들은 카피라이터들을 통해 다시 쓰여졌다. 사람들은 그들의 작업을 통해 중국식 빌바오 효과를 희망했으며, 지난 10년간 중국 역사를 돌아보면 건축가들은 학술적인 저널에서 상업적인 계약서로 옮겨갔고 마스터플랜은 실제 건물과 브랜드로 그 형태를 잡아갔다. 담론의 권리는 그 과정을 원만하게 통과한 사람들에게 돌아간다.

"행동"은 아마도 "열광"이라는 기치 하에서 일어나는 최고의 사건일 것이다. 그것은 또한 무서운 속도로 진행되는 중국의 개발 현실에 대한 본능적인 반응이거나 적극적인 선택일 수도 있다. 중국의 속도

는 임시 방편책을 요구하여, 새로운 세대는 '일단 하고 보자. 그리고 하면서 배우자'는 자세로 대응하고 있다. 젊은 세대는 일찍 철이 든 탓에 "늦게 철이 든" 이전 세대의 건축가들에게 세대차이를 느낀다. 하지만 진정으로 성숙해지기 위해서는, 권리나 부피보다 담론의 질적 향상이 더욱 중요하다. 마(Ma)의 세대가 독립적인 사고, 조직적인 실천, 광범위한 학제적 협력을 어떻게 발전시켜나가고, 직업적 명성과 이익을 어떻게 진정한 사업으로 탈바꿈시키며, 그들의 강도 높은 건축 작업을 어떻게 무게 있는 담론과 선언으로 이론화하는지, 그리고 마지막으로 열광과 이성을 어떻게 진보적인 건축적 이상으로 통합하는지를 지켜보는 것은 흥미로울 것이다.

On April 1, 2004, after finishing his studies abroad, Ma Yansong decided to hold an open ceremony on that April Fool's Day to introduce MAD - the architecture firm he'd registered in New York two years before - to China. The naming of architecture offices usually involves founders incorporating their fundamental ideas and beliefs as manifesto. Despite Ma's claim that MAD is nothing but an acronym of "MA-Design," the name's semantic suggestion does point to the firm's cynicism. Merely three years later, the name, along with Ma's head shot, became fixtures in many fashionable magazine cover stories. Ma, whose works are as jaw-dropping as Hollywood blockbusters, was called the model of the newer generation of Chinese architects, and his mode of success is now called the "Ma YanSong Axiom". How should one explain MAD's overnight fame and stardom? What has been the driving force behind the mutation from a simple office name to a popular brand name? Why is MAD mad?

To understand the relation between a Chinese architect and the social background of his generation, it would be helpful to break the Ma Yansong Axiom apart into elements of timing, place, and social network. The first element would be "the Chineseness card," namely, the advantage of being a Chinese architect in an international architectural context. The founding of MAD coincided with the crucial period of China's shifting towards a half-developed, modernized country. In spatial terms, this process of modernization manifested itself as large-scale, high-speed urbanization, thus creating a huge market for the building industry and a great number of architects.

On the other hand, with the enlightening, post-Open Door Policy "Great Leap Outward" period coming to an end, generations of culturally self-conscious Chinese prevailed. In the field of architecture, we witnessed the emerging of a number of creative architects. A market worth tens of billions of yuan annually, incredibly high design fees, and the personal fulfillment of participating in the magnificent process of urbanization in China are among the factors that make architecture an ideal profession for both fame and profit. The advent of the "starchitect's era" was simply a reflection of the zeitgeist. Meanwhile, the world was gazing at China curiously, wondering what kind of "stars" would rise from the drastically changing Middle Kingdom, how their perspectives and manifestos would redefine China in a global context, and what kind of changes they would bring to the country.

One card that goes hand-in-hand with the Chineseness card is "the international card". Coming back to China with a degree from a prestigious university in the United States, the license of a New York-registered office, and intern experience at top-notch international firms, Ma put together his own firm with an equally international lineup of principals, and became an aspiring participant in a series of international competitions and exhibitions. This strategy paid off in 2006, when MAD won the bid for the Absolute Tower in Toronto (better known as the "Marilyn Monroe building"), an event that put the firm on the way to stardom. As it turned out, Ma's decision to return to China before making his exploration outward was a natural fit with China's strategy of importing/internalizing foreign expertise so as to strengthen its presence on the world stage, a strategy known in the country as "bring-in-and-go-out."

During the same period, two Chinese architects of the previous generation (and with similar overseas educational backgrounds), Chang Yungho and Ma Qingyun, took the helm of the architectural schools in two of the most famous universities in the United States. The Monroe building, however, marks the first instance of a Chinese architect extending business overseas through major international competition. Driven by a nationalist complex, the domestic press interpreted the story as a kind of rise-of-the-nation event a la Yao Ming's entering the NBA, or Liu Xiang's world record-breaking performance. To be sure, undergoing the process of modernization, China needs exactly this kind of monumental events to mark its rising and national renaissance.

The Chineseness card interests the world, while the international card excites China: this logic of a "besieged fortress," where people trapped within want to go out and vice versa, exemplifies the un-saturation of China's modernity. It is also this logic that enables dual-background architects such as Chang Yungho, Ma Qingyun, and Ma Yansong to negotiate in both a domestic and international context. This kind of negotiation, in fact, is going to be the collective positioning of the next generation of free-thinking Chinese architects, a generation, according to Ma, whose voices will be collectively heard in the coming round of students returning from overseas. Ma is trying to fashion himself the leading example of these architects, who have made the same "Journey to the West" as he did.

To understand the dual-role of the two cards dialectically is a prerequisite to, in Ma's words, obtain the right of discourse. Where would he obtain that right? What is he going to discourse on?

There are basically three kinds of architecture organizations in Ma's China: The first one is the state-owned architecture design house. Being a part of the public sector, they get the lion's share of the building market, and are responsible for the mainstream urban landscape of China. The second kind is made up of international offices, which had swarmed in after the opening of the Chinese building market and the breaking down of regional and industrial restrictions. Adhering to international standards, these firms have injected fresh blood into the urban building enterprise of China. Local private design offices comprise the third kind. A result of urban construction shifting from top-down planning to market-oriented competition in the post-1990s era, these private offices have gradually become crucial players in the field.

All three types of organizations have their own merits: the state-owned houses, relying on their resources (often based on family ties) accumulated during years of experiences within the governmental mechanism, will be able to extend their monopolies for a considerable period of time. These houses will be granted most of the government-related large projects, but their bureaucratic administration, mechanical value judgment, and GDP-oriented fast mode of production dictate their conservative nature and the banality of their works. Nevertheless, a couple of design houses, who had the power trickle down during the social and political transformation, managed to produce their own starchitects by establishing relatively independent studios.

International firms can be divided into corporate ones and star-driven ones. As with most of the multinational corporations, corporate design firms have abundant funds, technological maturity, and proper administration. Star-driven firms, on the other hand, thrive on progressive thinking. These foreign offices are able to translate the experiences of "been there, done that" in developed countries and a mind-opening approach to project analysis into a core competitive advantage, thus making all state-owned firms worry about their own lack thereof. But they are also crippled by the difficulty of penetrating the reality of China and of understanding the rules in the local architectural scene. As a result, they have to succumb to the local protective policies and cooperate with state-owned offices.

The situation of the domestic private firms is the more subtle of the three. Less powerful than state-owned houses and not as prestigious as the international firms, they nonetheless absorb the merits of the two, as many of the private firms' founders have hands-on experiences in both state-owned and foreign offices. Not only do they know the reality of China, they are also internationally savvy; not only are they able to grasp all the tacit rules on the domestic business and governmental level, but also smart enough to adopt the latest international concept-in-vogue into their practices. Although these private firms occupy a smaller market share than the other two kinds of architectural entities, this kind of double savviness promises the potential to yield some really influential, deep thinking, radical starchitects who will be equally at home with theory and practice. They are going to comprise the local force confronting foreign star-driven firms. In the larger picture, it's safe to predict that the day when each of these three types of firm gets an equal share of the pie is around the corner.

In terms of the right of discourse, however, we see the market share hierarchy turned upside down. Granted, critical architectural discourse itself is by no means mature in China, but most of the current voices apparently come from A-list private firms. Thanks to several recent major China-related exhibitions in and outside of China and publications featuring conversations with confrontational ideas, a small, but much talked about circle of Chinese starchitects is coming into shape.

It seems that the misalignment of theory and practice in the Chinese architectural scene can be interpreted as the confrontation of sublime idealism and short-sighted utilitarianism: architecture gets redeemed in discourse if it falls short in practice. It seems necessary to erect a couple of monuments for the "pioneers" and "martyrs" - in the architectural sense - in order to promote their discursive value in the society through mass media, and, by doing so, to take aim at monopolies based on family ties and a building market over-obsessed with profit.

This kind of elitist "New Architecture Movement" of China morphed into a "New Starchitect Movement" when house building became commercialized and the country started to experience a building fever. This is especially true in a concept-driven city like Beijing, where starchitects were easily able to find bedfellows among real estate developers who were dying for a new tag line. The starchitect's name and head shot were then seen on billboards and agreements, the acclaims he'd won borrowed by copy writers. People were looking forward to the Chinese version of the Bilbao Effect from their works. In a sense, the history of Chinese architecture in the past ten years can be summarized as architects moving away from academic journals to commercial agreements and turning master plans into actual buildings and brand names. Those who have smoothly gone through that transition are the ones that obtain the right of discourse.

"Action" is probably the best footnote under the banner of "madness". It's also the instinctive response to and active choice in the current breakneck speed of China's development. The Chinese speed asks for an ad hoc strategy, to which the newer generation responds with a just-do-it-and-learn-in-the-process approach. It seems that their early maturity is bound to create a generation gap between them and the previous generation of "late-harvest" architects. The real maturity, however, depends not on the right or volume, but the quality of the discourse. It would be interesting to observe how Ma's generation is going to develop in terms of independent thinking, organizational practicing, and extensive interdisciplinary cooperation, how they are going to turn this profession of fame and profit into a real enterprise, theorize their intensive architectural practices into weighty discourse and manifesto, and, finally, fuse madness and reason into a progressive architectural ideal.

지앙 쥔은 디자이너이자 평론가다. 2003년 광저우에 언더라인 오피스(Underline Office)를 설립했으며, 2004년부터 어반 차이나 매거진의 책임 편집자로 활동하고 있다. 또한 그는 광저우 미술 아카데미에서 강의한다.
Jiang Jun is a designer and critic. He founded Underline Office in Guangzhou in 2003 and has been the editor-in-chief of Urban China Magazine since the end of 2004. He also teaches at the Guangzhou Academy of Fine Arts.

+ARCHITECT 02

STUDIO Pei-Zhu
STUDIO MAD

Inside Beijing Now

초판 1쇄 인쇄 2008년 9월 29일
초판 1쇄 발행 2008년 10월 1일
First Published in October, 2008

펴낸곳 (주)공간사
주소 110-280 서울시 종로구 원서동 219
출판등록 1978년 4월 25일 제1-18호
SPACE Publishing Co.
219 Wonseo-dong, Jongno-gu, Seoul, Korea
Tel +82-2-747-2892
Fax +82-2-747-2894
E-mail: editorial@vmspace.com
Homepage: www.vmspace.com

Publisher 이상림 Lee Sang-leem
Editorial Director 박성태 Park Seong-tae
Edition Editor 이정옥 Lee Jung-ouk
Assistant Editor 이지영 Lee Jee-young
Designer 박지혜 Park Ji-hye
Art Director 신범식 Shin Bum-shik
Marketing 이승연, 한경화 Lee Seung-yun, Han Kyoung-hwa

이 책의 저작권은 (주)공간사에게 있으며 저작권법에 의해 보호를 받는 저작물이므로 무단전재나 복제, 광전자 매체 수록 등을 금합니다.
All rights reserved. No part of this publication may be reproduced, stored in a retrieval system, or transmitted by any means, electronic, photocopying, recording, or otherwise, without the prior permission of the copyright owner.

Copyright ©2008 SPACE Publishing Co.

ISBN NO. : 978-89-85127-33-2 03600
20,000 won